GEOTECHNICAL PRACTICE PUBLICATION NO.

GEOTECHNICAL APPLICATIONS FOR TRANSPORTATION INFRASTRUCTURE

Featuring the Marquette Interchange Project in Milwaukee, Wisconsin

PROCEEDINGS OF THE 13TH GREAT LAKES GEOTECHNICAL AND GEOENVIRONMENTAL CONFERENCE

May 13, 2005
Milwaukee, Wisconsin

SPONSORED BY
University of Wisconsin-Milwaukee
Wisconsin Department of Transportation
Wisconsin Highway Research Program

CO-SPONSORED BY
The Geo-Institute of the American Society of Civil Engineers
American Society of Civil Engineers – Wisconsin Section
Milwaukee Transportation Partners
Edward E. Gillen Company
Center for Urban Transportation Studies of UW-Milwaukee
Midwest Regional University Transportation Center

EDITED BY
Hani H. Titi

Published by the American Society of Civil Engineers

Cataloging-in-Publication Data on file with the Library of Congress.

American Society of Civil Engineers
1801 Alexander Bell Drive
Reston, Virginia, 20191-4400

www.pubs.asce.org

ISBN 0-7844-0821-1
Manufactured in the United States of America.

Preface

This Proceedings contains 12 papers presented at the 13th Great Lakes Geotechnical and Geoenvironmental Conference (GLGGC) held in Milwaukee, Wisconsin, May 2005. Since major transportation infrastructure reconstruction was taking place in Milwaukee at the time of the conference, the conference theme addressed the role of geotechnical engineering in transportation infrastructure. The $810 million Marquette Interchange Project in the heart of downtown Milwaukee connects interstate highways I-43, I-94, and I-794. The role of geotechnical engineering in such projects is significant. The conference was designed as a forum for discussion of the latest advances in geotechnical engineering applications for transportation infrastructure including earth retaining structures, deep foundations, and geotechnical /pavement engineering.

Each paper published in this Proceedings was peer reviewed by at least one reviewer before being accepted for publication. The peer review was conducted in accordance with the Geo-Institute standards and practices. All papers are eligible for discussion in the Journal of Geotechnical and Geoenvironmental Engineering and for ASCE award nomination.

The peer review of the papers was conducted by the following reviewers: Murad Abu-Farsakh, Khalid Alshibili, Tuncer Edil, Mohammed Elias, Richard Finno, Dante Fratta, Mohammed Gabr, Sam Helwany, Andrew Heydinger, Baoshan Huang, Robert Liang, Eyad Massad, Radhey Sharma, Hani Titi, and Xiong (Bill) Yu.

I would like to express my appreciation to Nina McLawhorn of the Wisconsin Department of Transportation who made this work possible. I would like to acknowledge the support of the following: ASCE Wisconsin Section, HNTB, CH2M HILL, Edward E. Gillen Company, and Professor Edward Beimborn. I would like to thank John Siwula for his help and support in planning and organizing various activities. I would also like to thank the Technical Session Chairs: Nina McLawhorn, Bruce Pfister, Bob Arndorfer, Steve Maxwell, John Siwula, and Therese Koutnik.

The 13th GLGGC keynote speakers are Mr. Frank Busalacchi, Secretary, Wisconsin DOT and Mr. Jerry A. DiMaggio, principal geotechnical engineer and national program manager, FHWA. Their contribution to the conference is highly appreciated.

I would like to thank the authors and presenters of the papers for their contribution to the GLGGC. Special thanks are due to the following University of Wisconsin-Milwaukee individuals for their help in various conference activities: Jean Carlson, Mohammed Elias, and Dan Mielke.

Finally, I would like to acknowledge Harold Olsen and Richard Wiltshire for their help and assistance, the Geo-Institute Technical Publication Committee, Carol Bowers, and Donna Dickert.

Hani H. Titi, Ph.D., P.E.
Associate Professor, University of Wisconsin – Milwaukee

History of the Great Lakes Geotechnical and Geoenvironmental Conferences

The Great Lakes region is a highly populous and industrialized area of North America with a distinct economic and social character. The region was also subjected to active glaciation during the Pleistocene Epoch. This left behind unique surface and subsurface soil conditions and difficult groundwater problems. Because of this, the region has always posed great challenges to Geotechnical Engineers. These challenges continue to produce advances in classical Soil Mechanics, Foundation Engineering, and Groundwater Engineering. In addition, as the region struggles with environmental quality, the specialty of Geoenvironmental Engineering has matured to address problems such as groundwater pollution and solid waste disposal.

With these thoughts in mind, the Steering Committee of the annual Great Lakes Geotechnical/Geoenvironmental Conference has endeavored to ensure the continuation of the annual meetings. Each year a different theme is selected to focus on a topic pertinent to the Great Lakes region. Universities, local sections of ASCE or other professional organizations throughout the region host the conferences. It is hoped that these opportunities to discuss technical issues will be of value to researchers and practitioners dealing with problems related to soils of glacial origin. It is also hoped that the one-day format including oral and poster presentations, displays, and published proceedings will help facilitate the exchange of information among all of us struggling with the geotechnical/geoenvironmenal engineering challenges of the Great Lakes region.

Conference History

The Great Lakes Geotechnical and Geoenvironmental Engineering Conference is a one-day conference covering topics pertinent to the Great Lakes region. The purpose of the conference is to disseminate the latest information on current practices in engineering and related professions. The conference includes a Proceedings, including all conference papers presented at the conference. Previous and future Great Lakes Geotechnical and Geoenvironmental Engineering Conferences are listed in Table 1.

Awards

The GLGGC Steering Committee, responsible for organizing the conferences, meets on the evening of each conference. The GLGGC Steering Committee voted to establish awards for service, education and outstanding engineering at the 2000 meeting. Recipients of awards are usually recognized at the Annual GLGGC in the

Table 1: History and contact information of the Great Lakes Geotechnical and Geoenvironmental Conferences

Conf.	Title	Location	Date	Contact Person
1st	Engineering of Landfills in Glacial Soils	Toledo, OH	May 14, 1993	Andrew Heydinger, andrew.heydinger@utoledo.edu
2nd	Waste Materials and Their Geotechnical/ Geoenvironmental Applications	West Lafayette, IN	May 20, 1994	Rodrigo Salgado, salgado@purdue.edu
3rd	Geotechnical and Geoenvironmental Applications of Geosynthetics	Cleveland, OH	May 19, 1995	Andrew Heydinger, andrew.heydinger@utoledo.edu
4th	Insitu Remediation of Contaminated Sites	Chicago, IL	May 17, 1996	Krishna R. Reddy, kreddy@uic.edu
5th	Site Characterization for Geotechnical and Geoenvironmental Problems	Ann Arbor, MI	May 16, 1997	Richard Woods, rdw@engin.umich.edu
6th	The Design and Construction of Drilled Deep Foundations	Indianapolis, IN	May 8, 1998	Andrew Heydinger, andrew.heydinger@utoledo.edu
7th	Recent Advances in Ground Improvement Methods	Kent, OH	May 10, 1999	Andrew Heydinger, andrew.heydinger@utoledo.edu
8th	Geotechnology for Urban Renewal and Redevelopment	Detroit, MI	May 19, 2000	Nazli Yesiller, yesiller@eng.wayne.edu
9th	SlipSlidin' Away Contemporary Solutions to Land Mass Stabilization	Dayton, OH	May 11, 2001	Manoochehr Zoghi, Manoochehr.Zoghi@notes.udayton.edu
10th	Transportation Geotechniques	Toledo, OH	May 10, 2002	Andrew Heydinger, andrew.heydinger@utoledo.edu
11th	Advances in Characterizing and Engineering Problem Soils	West Lafayette, IN	May 23, 2003	Maria Santagata, mks@purdue.edu
12th	Advances in Deep Foundation, Design Construction and Quality Control	Akron, OH	May 7, 2004	Robert Liang, rliang@uakron.edu
13th	Geotechnical Applications for Transportation Infrastructure	Milwaukee, WI	May 15, 2005	Hani Titi, hanititi@uwm.edu
14th	Earth Retention/Coastline Protection Design and Construction	Chicago, IL	May, 2006	Sara Knight, knight@stsconsultants.com

year following their selection. Table 2 presents information about the individuals who have been recognized by the committee.

Table 2: GLGGC Awards

Award	Recipient	Affiliation	Year Award
Excellence for Service	Bill Lovell	Purdue University	2001
Distinguished Engineer	Richard Woods	University of Michigan	2001
Excellence Plaque	Bill Lovell	Purdue University	2001
Excellence Plaque	Andrew Bodosci	University of Cincinnati	2001
Excellence Plaque	Donald Gray	University of Michigan	2001
Excellence Plaque	Robert Lennertz		2001
Distinguished Engineer	Vincent Drnevich	Purdue University	2002
Special Service Award	Andrew Heydinger	University of Toledo	2002
Excellence for Service	William Cutter	ATC Associates, Inc., Indianapolis, IN	2003
Excellence for Geotechnical Education	Brian W. Randolph	University of Toledo	2003
Outstanding Educator in the Region	Manoochehr Zoghi	University of Dayton	2004
Distinguished Geotechnical Engineer in the Region	Clyde Baker	STS Consultants, Ltd, Vernon Hills, Illinois	2005
Excellence in Service	Robert Liang	University of Akron	2005
Outstanding Geotechnical Educator in the Region	Kevin Sutterer	Rose-Hulman Institute of Technology	2005
Distinguished Geotechnical Engineer in the Region	John Siwula	HNTB, Milwaukee	2006
Excellence in Service to GLGGC	Hani Titi	University of Wisconsin - Milwaukee	2006
Outstanding Geotechnical Educator in the Region	Krishna Reddy	University of Illinois - Chicago	2006

13th Annual Great Lakes Geotechnical and Geoenvironmental Conference

The 13th GLGGC was held at the University of Wisconsin-Milwaukee Campus in Milwaukee on May 13, 2005. The conference was organized by the Department of Civil Engineering and Mechanics, University of Wisconsin-Milwaukee and Wisconsin Department of Transportation/Wisconsin Highway Research Program. The conference planning committee consisted of the following individuals:

Hani Titi, Sam Helwany, and Mohammed Elias – University of Wisconsin-Milwaukee
Nina McLawhorn, Bruce Pfister, and Robert Arndorfer – Wisconsin Department of Transportation

John Siwula, Brian Powell, Therese Koutnik, and Viki Streich – Milwaukee Transportation Partners
Tuncer Edil – University of Wisconsin-Madison
Alan Wagner – Wagner Komurka Geotechnical Group
Timothy Bate – American Society of Civil Engineers, Wisconsin Section

Planning for the 13th GLGGC started on May 2004 with the steering committee meeting. In the 13th conference, the steering committee has chosen to address the role of geotechnical engineering in transportation infrastructure. The conference is intended to be a forum for discussion of the latest advances in geotechnical engineering applications for transportation infrastructure including earth retaining structures, deep foundations, and geotechnical /pavement engineering. The conference will feature the Marquette Interchange Project in Milwaukee. Based on that the following conference themes were adopted:

- Innovative construction methodologies of earth retaining structures
- Testing/instrumentation of earth retaining structures
- Innovative construction/design of deep foundations
- Innovative geotechnical applications in pavements
- Case histories

The 2005 GLGGC Steering Committee consists of the following individuals:

President: Robert Liang – University of Akron
Vice President: Hani Titi – University of Wisconsin-Milwaukee
Secretary/Treasurer: Krishna Reddy – University of Illinois
Executive Director: Andrew Heydinger – University of Toledo
Member Emeritus: C.W. "Bill" Lovell – Purdue University
Member Emeritus: Richard D. Woods – University of Michigan
Member: Marika Santagata – Purdue University
Member: Manoochehr Zoghi – University of Dayton
Member: Radhey Sharma, Iowa State University
Member: Sarah Knight: STS Consultants

A website was established for the conference detailing information needed. The website address is: http://www.uwm.edu/CEAS/glggc/. In addition, video of the 13th GLGGC presentations is available on the following website, http://www.dot.wisconsin.gov/library/research/reports/whrp.htm.

Contents

AUTOMATED MONITORING OF SUPPORTED EXCAVATIONS

Richard J. Finno[1], M.ASCE and J. Tanner Blackburn[2]

ABSTRACT: This paper describes remote and automatic monitoring systems for measuring vertical and horizontal displacements and tilt of structural elements. An automated total station and system of tiltmeters were employed at the excavation for the Ford Engineering Design Center. The 9.1-m-deep excavation was made through fill and soft clays and undercut the shallow foundations for a building 6 m from the edge of the excavation. The excavation support system consisted of a sheet pile wall supported by two levels of cross-lot and diagonal internal bracing. The two automated systems, the structural support system and construction details are described. Selected ground and structural responses collected by the automated systems are compared to those collected with conventional methods. The accuracies of the measured automated responses are discussed. The tilt meter data are shown to supplement the inclinometer data so that structural responses of an adjacent structure to the excavation can be evaluated.

INTRODUCTION

A number of performance data for deep supported excavations can be remotely and automatically monitored as a result of advances in data acquisition and sensing systems. A commercially available total survey station was modified to robotically monitor a series of stationary prisms attached at various points around the excavation for the Ford Motor Company Engineering Design Center (FEDC) on the campus of Northwestern University in Evanston, IL. Several tiltmeter pairs were installed on structural elements of the adjacent Technological Institute (Tech Building) to observe building deformation during the excavation. The data were retrieved remotely and automatically, via radio link and processed on a central data server.

The FEDC building includes two basement levels, requiring an excavation of 9.1 m. The excavation was supported by sheet pile walls and two levels of interior bracing. The excavation was adjacent to the Technological Building which is founded on strip and spread footings. Because of the close proximity of the Tech Building, soil and structural responses were monitored by two automated systems, an optical survey station to measure ground surface and sheet-pile wall deformations and tiltmeters affixed to several columns in Tech, as well as inclinometers on 3 sides of the excavation and strain gages applied to several support struts.

---

[1] Professor of Civil Engineering, Northwestern University, Evanston, IL 60208, r finno@northwestern.edu

[2] Research Assistant, Northwestern University, Evanston, IL 60208, j-blackburn1@northwestern.edu

This paper describes the two automated monitoring systems and summarizes typical responses recorded by the systems. A web camera was also used to allow observations of the construction in real time at remote locations. Accuracy of the total station is shown under weather conditions typical of a Midwest winter. Comparisons are made between the remotely collected data and similar data collected with conventional means.

CONDITIONS AT THE FEDC

A plan of the excavation for the FEDC is presented in Figure 1 showing the internal support, the adjacent Tech Building and locations of the automated monitoring points. The excavation measured approximately 44.5 m by 36.6 m and was supported by a sheet pile wall consisting of XZ85 section and two levels of internal bracing. The excavation was 9.1 m deep from the original grade at the center of the site and reached a final grade of elev. -3.8 m Evanston City Datum (ECD). W24x141and W36x230 walers were attached to sheet piles at the first and second level of bracing, respectively. Pipe struts with diameters of 0.61 m were used for the cross lot bracing and were supported in the center by H-pile vertical and lateral supports. Three diagonal supports were placed in each corner; pipe struts were used for the two longer sections and a W14x145 section was used for the section nearest the corner. Tech was located within 5 m of the FEDC excavation and is supported by shallow footings at elevations between 0.4 and 1.5 m ECD.

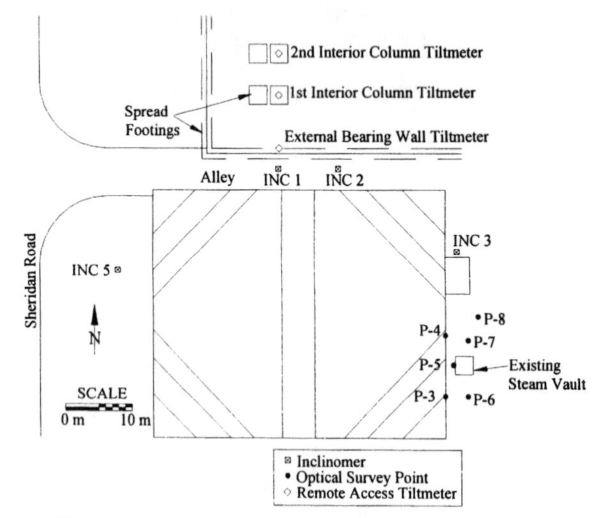

FIG. 1. Ford Engineering Design Center Excavation

Figure 2 shows a section through the excavation, the soil stratigraphy and water content and undrained shear strength profiles. A stiff clay crust with variable thickness was encountered below the sandy fill material. The softer clays at the site are ice margin deposits with an undrained shear strength, which generally increases with depth. The hardpan is a basal till that is much stiffer and stronger than the overlying clay strata. Chung and Finno (1992) provide a more complete description of the local geology.

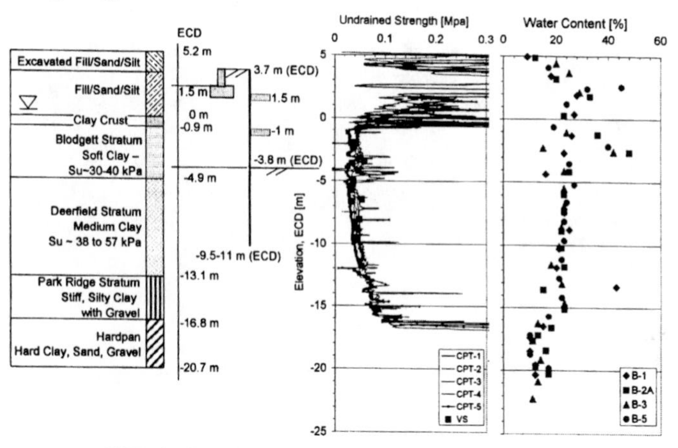

FIG. 2. Subsurface conditions at the FEDC site

AUTOMATED MONITORING SYSTEMS

Total Surveying Station

A Leica TPS1101 Professional total surveying station was used to monitor the displacement of optical prisms placed throughout the FEDC site. The total station is shown mounted via an aluminum mounting plate and swing-arm apparatus on the roof of the Tech Building in Figure 3. Once the total station was secured into place, the instrument was leveled and initialized for remote monitoring. The instrument includes monitoring firmware and robotic, automatic target recognition capabilities (ATR), which enabled repeated automated position measurement of predetermined points surrounding the excavation. To utilize the ATR capabilities, a user first manually locates a series of points marked with prisms.

Optical surveying prisms were installed at selected locations to monitor the deformation of adjacent soil and structures during the excavation process. Figure 1 shows the locations of the monitoring points (P series) adjacent to the FEDC excavation. Additionally, two prisms were mounted on structures located over 60 m from the excavation to serve as stationary reference points (P-1 and P-2) so that the settlement of the other points could be calculated. Two prisms (P-3 and P-4) were mounted to the east sheet-pile wall of the FEDC retention system to monitor outward, cantilever deflection and sheet-pile settlement. Three prisms were attached to threaded rods that were embedded into the soil with concrete in the soil adjacent to the east side of the FEDC excavation. An additional prism (P-5) was attached to a steam vault directly adjacent to the sheeting.

FIG 3. Total Station on rooftop of Tech Building.

A series of Matlab® scripts were written to communicate with the total station at regular time intervals, collect point displacement data, translate and parse data, and process data into settlement plots and files. The total station was configured to allow communication via RS232 data transmission. Data collection and remote operation errors can occur as a result of instrument power loss, communication errors, obscured survey prisms, and instrument disturbance. These errors are detected with the Matlab® scripts and documented in the log file

All point position data are calculated relative to the two reference points. Calculating relative positions based on the remote reference points, rather than assuming that the total station is fixed, allows for slight instrument movement during operation and allows the instrument to be removed for repairs, if necessary. This is important when the total station is exposed to the rigors of a construction site for an appreciable period of time. Also, by calculating the positions of the reference points relative to each other, instrument tilt can be detected. If a significant change in position of P-2 or P-1 is detected, the instrument is out of level, or one of the two reference points has been disturbed.

Remote Access Tiltmeters

Ten pairs of single-axis tiltmeters were installed on selected structural elements of the Tech Building to monitor the response of the structure during excavation. The tiltmeter models are Schaevitz Accustar Electronic Clinometers, both analog and ratiometric versions were used for the FEDC project. A system of commercially available analog-to-digital converters and data signal converters (CyberResearch, CyMOD 4520 and 4017) was used to convert the voltage output to a digital value that was then retrieved using radio communications.

Figure 1 shows the general locations of the tiltmeter pairs whose responses are summarized herein. Four tiltmeter pairs were installed in the basement of the Tech Building, on supporting columns or walls, just above the top of the spread footings, which have a base elevation of approximately +1.5 m ECD. Three of the basement pairs were installed along a column line perpendicular to the excavation, which would detect differences in building tilt as a result of foundation movement. An additional pair was installed on the west wall of the Tech Building to distinguish the different tilt patterns due to distortions that develop parallel to the north side of the excavation.

The tiltmeters were attached to the described columns and walls of the Tech Building with a quick-setting epoxy and coupled with Freewave point-to-point radio transceivers to allow continuous remote data collection. Figure 4 gives a photograph of a tiltmeter pair, affixed with epoxy to a column in the basement of the Tech Building.

Several Matlab® scripts were written for periodic remote data collection from all of the tiltmeters. Individual voltage values are read fifty consecutive times and then averaged, which effectively reduces effects of electronic noise on the system. The individual averaged voltage is converted to a tilt value for each inclinometer and added to a text file, which contains a time history of collected data for each area. An additional Matlab® script is used to specify the frequency and schedule of data collection.

FIG. 4. Photograph of tilt meter installed in the Tech Building basement

Remote Communication Methods

The RS232 serial port communication mode of the total station allowed for several methods of remote communication between the total station and the data collection server. For this project traditional modem communication, cellular phone modem communication and point-to-point radio communication were explored. Matlab® scripts were written for each communication option.

Table 1 lists the three types of remote communication and the advantages and requirements of each method. Traditional modem communication has a lower initial cost and reliable data transfer, and is the best current solution for urban environments where telephone lines are common and convenient. Cellular modem communication can be employed at more remote sites, where traditional telephone lines are rare. Radio point-to-point communication can be employed when the data collection server is within several miles of the site, and there are few obstructions between the server and the total station. Direct RS232 to Ethernet data conversion is also listed in Table 1 because it is a viable option in urban environments; however, it was not explored for the FEDC project. A combination of communication methods can also be used to access data remotely. For example, a radio point-to-point system can transmit data to a computer in a construction trailer, which then can be accessed via internet or modem by a remote user. All of the methods mentioned above require a power supply; however, deep cycle batteries can be used, which require charging or replacing on a monthly basis depending on data collection frequency.

TABLE 1. Communication methods for remote monitoring

Communication Method	Advantages	Requirements
Traditional Modem	Limitless range Availability of phone lines in urban environments Low initial costs	Dedicated telephone line AC power or deep cycle battery
Cellular Telephone Modem	No need for phone line	AC power or deep cycle battery Cellular service in region
Spread Spectrum, Point-to-Point Radios	No phone line necessary No cellular service necessary	AC power Few obstructions between transceivers
Ethernet	Direct data transfer to PC Accessible by many PCs Reliable data transfer	Ethernet port at instrument location AC power

A system of point-to-point, spread spectrum, radio transmission was employed for communication between the data collection server and total station at this site. The radio transceivers were commercially manufactured (Freewave, DGR -115R) and have a range of several miles, depending on the number and types of obstructions between transceivers. One transceiver is connected to the total station by 9-pin serial connection and communicates with the total station via RS232 transmission. This transceiver placed on the roof of Tech (Figure 3) also serves as a relay point for additional radio data transmission from the tiltmeters, which are discussed subsequently. An additional transceiver is located at the data collection server, which communicates with the Roof Transceiver and the tiltmeters.

Internet Accessible Video Camera

An internet accessible weather-resistant video camera ("webcam") was installed on the roof of Tech to monitor the FEDC construction process in real-time, and is shown in Figure 3. The images were accessible via the Northwestern University website to the general public, including project engineers and contractors. The webcam images were combined with the automated survey and tiltmeter data to provide an opportunity to observe the response to the excavation process in real-time from a remote location and to relate it qualitatively to construction operations.

SURFACE SETTLEMENT

The automated, remote-access, optical survey system monitored vertical settlement and horizontal displacement at selected locations (Figure 1). The survey point locations were automatically recorded on a frequent basis during the excavation. Because of the exposed location of the instrument and instability of the mounting apparatus, winds would cause the instrument to go out of level on occasion. An internal mechanism would halt operation if the instrument exceeded a specific 'out-of-level' threshold. Therefore, there were periodic gaps in the data collection, where the instrument would be out of operation until the instrument was reset. The instrument was operational during extreme winter conditions, including a low temperature of -5° C. Snowy conditions did not affect the instrument operations; however snow occasionally obscured the soil-mounted prisms.

Survey System Accuracy

Two benchmark points mounted several meters above ground surface on permanent structures 65 m away from the excavation provided references from which movements of the excavation monitoring points were computed. The relative displacement between these two benchmarks (P-1 and P-2) should be zero and can be used to assess the accuracy of the monitoring system. Figure 5 is a plot of the relative movement between these 2 points and reflects the accuracy of the measurements. This variability is attributable to instrument movement due to wind and weather, and thus must be accounted for in surface settlement monitoring. The standard deviation, σ, of the benchmark settlement (from zero) is 1.8 mm. Approximately 94% of the data points fall within 2σ of the mean, or 3.6 mm, and approximately 99% of the data points fall within 3σ of the mean, or 5.4 mm. Thus, it is reasonable to estimate the accuracy to within 2σ for this project as 3.6 mm. This is slightly higher than the 3.2 mm accuracy claimed by traditional optical survey professionals.

To increase the accuracy of these measurements, a 10-point moving average was employed for tracking the survey point position. According to product documentation, the total station accuracy during automated measurement is 3 mm at distances up to 300 meters, at which the accuracy corresponds to the angular accuracy of 1.5 seconds, or approximately 7.3 parts per million. Figure 6 shows the 10-point moving average of the relative vertical movement between the benchmarks. The standard deviation of the 10-point moving average data is 0.8 mm, which is a significant improvement over the 1.8 mm standard deviation without averaging. For data recorded throughout the project and processed with the 10-point moving average, the 3σ accuracy is 2.8 mm, within the stated accuracy given by the manufacturer.

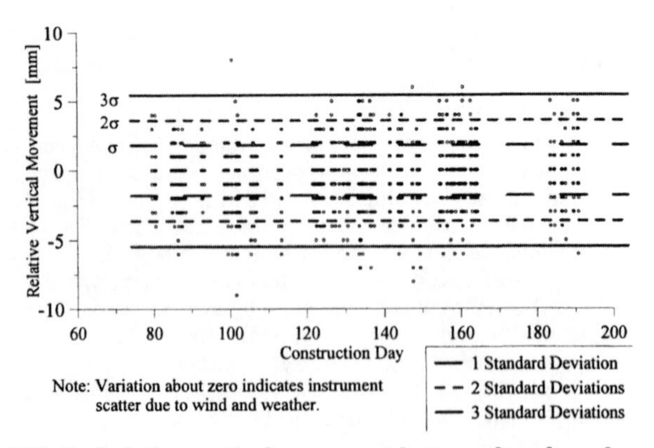

FIG. 5. Relative vertical movement between benchmarks.

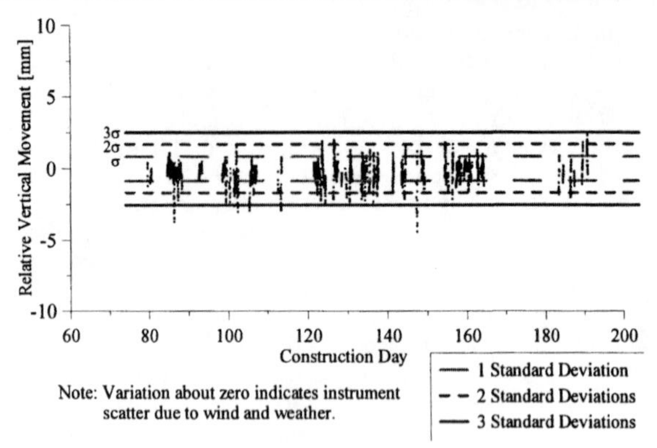

FIG. 6. 10-Point moving average of relative vertical movement between benchmarks

Displacements of Soil and Steam Vault Points

Settlement of the three soil survey points (P-6,7,8) and the survey point mounted on the concrete steam vault directly adjacent to the excavation was monitored throughout excavation, and are shown in Figure 7. The steam vault settled approximately 8 cm during the excavation. Soil points P-6, P-7, and P-8 settled approximately 5 cm, 7.5 cm, and 4.5 cm, respectively, during the excavation process. Large settlement increments occurred as the southeast corner was excavated from elev. +1 m to -1.5 m ECD and from -1.5 m to -3.8 m ECD. The steam vault was also monitored by a local surveying company, using a traditional optical survey, on a monthly basis. The settlement reported by the surveyor corresponds to the automated survey data until construction day 130. However, the surveyor reported no additional settlement beyond this date, which contradicts the data recorded by the automated survey system. Given the correlation of the movements with the depth of excavation in front of the wall, it is likely the conventional survey data were incorrect after day 130. These date were transmitted to the contractor 2 weeks after the data were recorded, in contrast to the real time acquisition possible with the automated system.

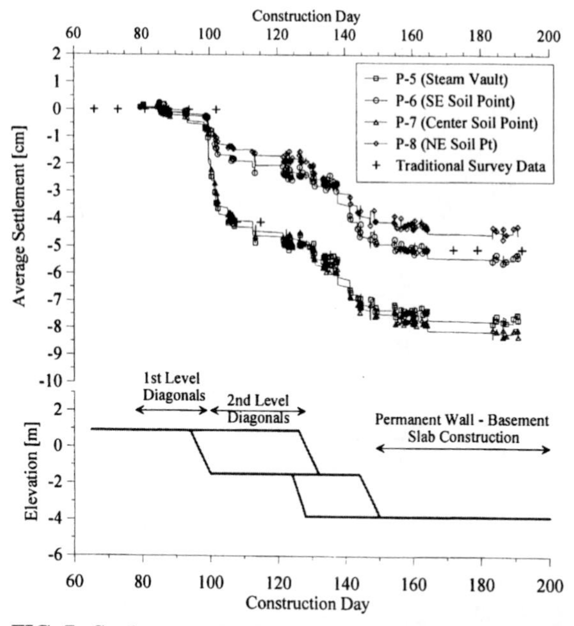

FIG. 7. Settlement of soil and steam vault survey points

Figure 8 is given to compare the lateral soil displacement measured by Inclinometer 3 on the east side of the excavation with the surface displacements observed with the automated survey system. Lateral movements from inclinometer data are shown from two depths, elev. +1.8 m and -4.9 m ECD. These depths reflect the cantilever displacement at the surface (1.8 m ECD) and the deep-seated deformation in the soft clay (-4.9 m ECD). The survey data plotted on the figure is the 10-point moving average value corresponding to the inclinometer reading date. As expected, the initial soil settlement and eastward movement increases coincide with a cantilever movement increase, measured by Inclinometer 3. Also, the second jump in soil settlement and eastward movement coincides with an increase in movement toward the excavation in the soft clay layer (construction day 127).

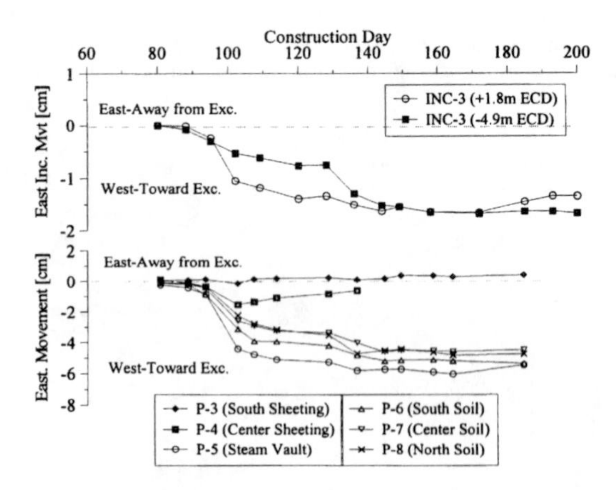

FIG. 8. Soil displacements measured remotely and by Inclinometer 3 at two depths

ADJACENT BUILDING RESPONSES TO EXCAVATION

Remote access tiltmeter pairs were installed on support members at various locations of the Tech Building, including in the basement of the 4-story section immediately adjacent to the excavation. The Tech Building spread footings are located at a range of elevations, from +0.4m to +1.7m ECD, which is in the lower region of the sand and fill layers of the FEDC stratigraphy. If the deformation in the underlying clay layers is assumed to be undrained (with zero volume change), then the vertical settlement distribution at the footing elevation can be approximated by rotating the lateral movements from an inclinometer about a pivot point at a depth corresponding to the inclinometer distance from the wall (Finno, et al. 2002), as illustrated in Figure 9.

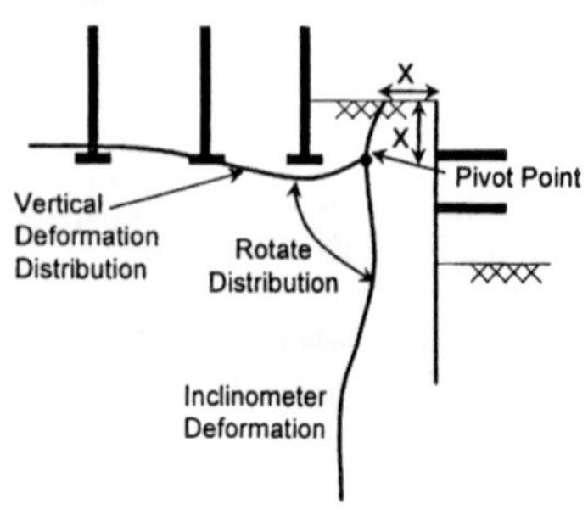

FIG. 9. Obtaining approximate soil settlement profile from inclinometer data

The slope of the footing at each column location is estimated from the rotation of the slope inclinometer deformation pattern. Tiltmeters installed on support columns just above the floor slab of the Tech Building recorded the column tilt as the footing deforms with the soil. When the tiltmeter results are compared with the expected slope at each footing location, the difference between the two is caused by restraint from the superstructure and the stiffness of the column-footing connection. For a rigid connection, the tilt equals the slope.

Figure 10 shows measured tilt values at the wall and first interior column in the north/south direction, the slope of the footing from inclinometer 1 data, the distortion between the wall and first interior column, and a brief construction record. The distortion is the relative settlement between the wall and the second column divided by the distance between the two and depends on the structural stiffness of the building.

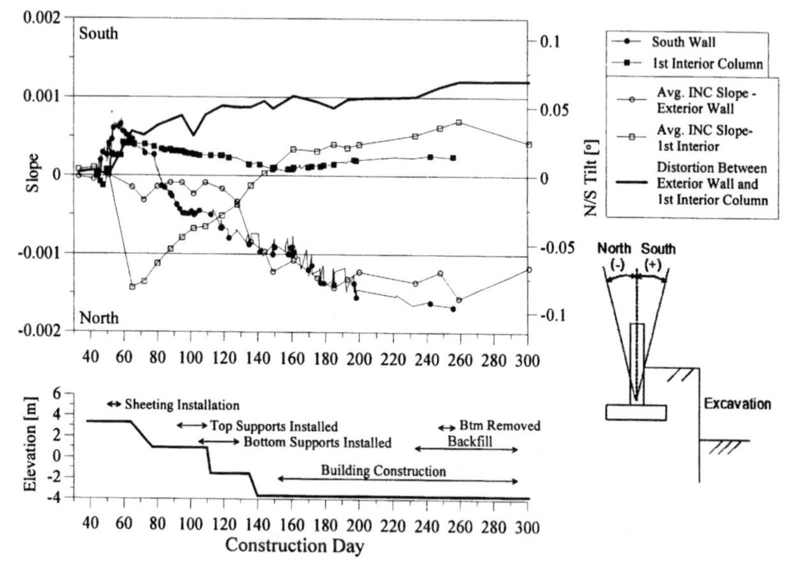

FIG. 10. Comparison of north/south basement tilt data, soil slope from inclinometer, and distortion between the 1st interior column and the exterior wall.

Initially, the clay soil heaves away from the excavation as a result of installing the sheeting around day 50, resulting in heave of the structure. This movement is more pronounced in the clay, so the maximum heave occurs at some distance from the south end of the building, resulting in more heave at the column. The distortion to the south, implying that the column heaved more than the wall, reflects this response, as does the tilt of both column and wall. As the excavation progresses and the soil moves toward the excavation, both the column and wall footing settle. The wall tilts towards the north at approximately the same rate as the soil slope from the inclinometer data, suggesting an almost rigid connection between the thick masonry wall and underlying strip footing. The magnitude of the tilt observed at his location approached a slope of 0.0015, or 1/667. Tilt of the second column does not track the slope of the inclinometer data at the corresponding depth between days 62 and 140, but rather more closely mimicked the distortion. This pattern of movement suggests a more flexible connection between the footing and the column, as one would expect for an isolated column on a spread footing as compared to a thick wall on a strip footing. After day 140 when the excavation had reached its final depth, the column tilt of less than 1/2000 was approximately equal to the slope of the footing. Despite these levels of column tilting and footing deformations, no structural or cosmetic damage to the Tech Building was detected in this direction during the excavation.

CONCLUSIONS

Soil deformation and adjacent structural response was monitored by several instrumentation methods during the FEDC excavation process. Automated optical surveying and tiltmeter systems successfully monitored in real time surface settlement adjacent to the Ford Center excavation and the tilt of structural elements of the adjacent structure. Slight tilting of the columns and walls were observed, on the order of 1/1000 to 1/500. However, these distortion levels did not result in any cosmetic or structural damage to the building.

ACKNOWLEDGEMENTS

This work was made possible by the cooperation of Thatcher Engineering, Inc., the excavation support subcontractor and Turner Construction, the general contractor at the Ford Center. The authors thank Dr. Michael Wysockey of Thatcher and Mr. Philip Blakeman of Turner for their help and interest in the work. The work was funded by funds from grants CMS-0084664 and CMS-115213 from the National Science Foundation (NSF). The junior author was supported partially by a grant from the Infrastructure Technology Institute (ITI) at Northwestern University. The authors thank Dr. Richard Fragaszy, program manager of Geomechanics and Geotechnical Systems at NSF and Mr. David Schulz, director of ITI, for their support.

REFERENCES

Chung, C.-K. and Finno, R.J. (1992), "Influence of Depositional Processes on the Geotechnical Parameters of Chicago Glacial Clays," *Engineering Geology*, 32, 225-242.

Finno, R.J., Bryson, S. and Calvello, M. (2002), "Performance of a Stiff Support System in Soft Clay," *J. of Geotech. and Geoenvir. Engrg.*, ASCE, 128 (8), 660-671.

THE USE OF IN-SITU TESTING TO OPTIMIZE RETAINING WALL DESIGN IN THE MARQUETTE INTERCHANGE PROJECT

Emad Farouz[1], P.E., Member ASCE, Jiun-Yih Chen[2], E.I.T., Member ASCE, Therese E. Koutnik[3], P.E., Member ASCE, and Brian L. Powell[4], E.I.T., Member ASCE

ABSTRACT: This paper presents the practical use of in-situ testing to optimize retaining wall design of the $810 Million Marquette Interchange Project in Milwaukee, Wisconsin. Approximately 42 retaining walls will be constructed between 2004 and 2008, of which the majority are cut walls up to 12.2-meter (40-foot) high. The in-situ testing performed for this project supplemented a conventional geotechnical investigation and included pressuremeter tests, dilatometer tests, and piezometric cone penetration tests with soil electrical conductivity measurements. Pressuremeter and dilatometer test results were used to develop categorized p-y curves in modeling passive soil resistance and to estimate undrained shear strengths and anchor bond strengths of clayey soils. Sounding results from cone penetration tests with soil electrical conductivity measurements were used to evaluate the groundwater conditions, estimate the in-situ soil properties, and perform direct settlement analyses of fill walls on soft ground conditions. In-situ testing results were compared with soil parameters developed from Standard Penetration Tests and laboratory tests. Results were also validated from three full-scale lateral pile load tests conducted during the design phase and from anchor load tests conducted as part of construction quality control/quality assurance. This paper illustrates how the in-situ testing results were incorporated to optimize the retaining wall design and reduce overall construction costs.

---

[1]Senior Geotechnical Engineer, CH2M HILL, Inc., 13921 Park Center Road, Suite 600, Herndon, VA 20171, Emad.Farouz@CH2M.com

[2]Geotechnical Engineer, CH2M HILL, Inc., 3921 Park Center Road, Suite 600, Herndon, VA 20171, Jiun-Yih.Chen@CH2M.com

[3]Geotechnical Engineer, HNTB Corporation, 11414 West Park Place, Suite 300, Milwaukee, WI 53224, tkoutnik@hntb.com

[4]Geotechnical Engineer, HNTB Corporation, 11414 West Park Place, Suite 300, Milwaukee, WI 53224, bpowell@hntb.com

INTRODUCTION AND BACKGROUND

The $810 Million Marquette Interchange Project is a reconstruction of an interchange located in downtown Milwaukee that connects 3 major interstate highways: I-43, I-94, and I-794. The Milwaukee Transportation Partners (MTP) was tasked with the engineering design of 42 retaining walls constructed between 2004 and 2008.

Retaining walls range in height from 1.2 to 12.2 meters (m) [4 to 40 feet (ft)] and are located in geology primarily comprised of either glacial till or estuarine deposited river valleys. Cut walls consisted of 55 percent of the total number of retaining walls. The remaining 45 percent of the walls were fill or cut/fill walls consisting of either mechanically stabilized earth (MSE) walls with overexcavation or cantilevered cast-in-place (CIPC) walls.

In-situ testing performed for this project supplemented a conventional geotechnical investigation and included pressuremeter tests (PMT), dilatometer tests (DMT), and piezometric cone penetration tests with soil electrical conductivity measurements (CPTU-EC). In-situ testing results were compared with soil parameters developed from the Standard Penetration Tests and laboratory tests and validated with the results from three full-scale lateral pile load tests conducted during the design phase.

This paper discusses the geology of the project; in-situ testing performed; the development of soil models for analyzing piles under lateral loading from the in-situ testing; and how the in-situ testing results optimized the retaining wall design to reduce overall construction costs.

GEOLOGY AND GENERAL SUBSURFACE CONDITIONS

The Quaternary geology of the project region is complex due to the varying depositional mechanisms and environments of glacial, glaciofluvial, lacustrine, alluvial, and estuarine origin. Historic development of the City of Milwaukee resulted in cutting local uplands and filling depressions and low wetland areas that existed near and adjacent to the river valleys. Much of the cut and fill activity occurred in the latter half of the 19th century. Consequently, a wide distribution of fill material exists at the surface in much of the project area that masks the upland till and valley soil deposits.

The principal regional physiographic features that encompass the Marquette Interchange Project include Valleys and Lowlands of the Menomonee River, Milwaukee River, and Lake Michigan shoreline; and Upland areas that rise sharply along the bluffs of the Valleys and Lowlands.

Silurian Racine Dolomite forms the bedrock surface throughout the majority of project region with few exceptions. The deepest depth to bedrock surface underlies the Menomonee River Valley at a depth of 45.7 to 73.2 m (150 to 240 ft) below grade surface.

IN-SITU TESTING AND SUBSURFACE INVESTIGATION PROGRAM

For the Marquette Interchange Project, the total geotechnical investigation program completed in 2002 and 2003 consisted of 405 borings; various conventional laboratory testing; 13 temporary wells, 15 permanent monitoring wells; 24 PMT at retaining wall locations; 27 CPTU-EC; and 2 DMT.

Laboratory testing included moisture content, particle size distribution, Atterberg limits, unit weight, specific gravity, unconfined compression on soil and rock, loss on ignition, one-dimensional consolidation, consolidated-undrained and unconsolidated-undrained triaxial tests, modified proctor, California Bearing Ratio, permeability, pH, resistivity, sulfates, and chlorides.

For the pressuremeter testing, three to five tests were performed at each location with depths up to 19.8 m (65 ft) below grade. The test locations were planned at cut walls and high-level ramp structures where foundation elements required higher soil resistance to lateral loads. Typically, the tested soil was described as fill, clayey silt / silty clay, or organic clay / silt soils and were representative of the subsurface conditions in this area. Results of PMT were used to develop the p-y curves and to estimate undrained shear strengths of clayey soils. P-y curves were used to model the non-linear soil resistance to the embedment of retaining walls and pile groups. The undrained shear strength of cohesive soils can be used to estimate bond strength of tieback anchors.

CPTU-EC testing was performed up to 28.2 m (92.6 ft) below grade. The testing was performed at locations, where groundwater levels were anticipated to be high and high fill walls would be constructed, to determine groundwater levels and to estimate settlement. Measurements of generated pore pressure during penetration and soil electrical conductivity indicated more accurately the groundwater and perched water levels than conventional borings without monitoring wells. The measured cone tip resistance, friction ratio, and pore water pressure was used to estimate undrained shear strength and overconsolidation ratio in clayey soils and drained friction angle in sandy soils. From the pore water dissipation tests, the coefficient of consolidation, hydraulic conductivity, and constrained modulus were determined and used to estimate long-term settlement from roadway embankments or fill walls.

The two DMT were performed to approximately 4.6 and 6.1 m (15.2 and 20.0 ft) below grade within an organic clay / silt material. DMT results were used to determine the soil strength, compressibility, and stress history. These results were used in the evaluation of settlement of fill walls near the test locations.

The PMT results were validated from three full-scale lateral pile load tests that were completed as part of the Design Phase Pile Load Test Program for the Core Bridges. At each test location, two closed-ended, cast-in-place, concrete-filled, steel pipe piles varying in size from 324 millimeters (mm) [12.75 –inches] to 406 mm (16 inches) were jacked apart to failure or up to 400 kilonewtons (kN) [45 tons]. Inclinometers measured the deflection of the pile versus depth, which was used to validate the p-y curves developed from the PMT. The test locations were selected where greater numbers of foundations were planned and where higher lateral bridge loads were anticipated. The soil profile in each location was also representative of one of the general soil profiles across the site.

RETAINING WALL TYPES

A variety of retaining walls were used in the Marquette Interchange Project as grade-separation structures. They were classified into two major categories: (1) cut walls, and (2) fill walls or cut/fill walls with minor overexcavation.

Cut Walls

Cut walls evaluated and selected for the Marquette Interchange included secant / tangent pile, sheet pile, soldier pile and lagging, and soil nail walls. Among the cut walls, secant pile walls were the majority as groundwater cutoff walls in this project. Because the cut was often below the groundwater table, using a groundwater cutoff wall was necessary to avoid lowering the groundwater table in the predominant clayey soil profiles and consequently inducing long-term settlement of existing structures and utilities behind the walls. Additionally, secant pile walls are relatively stiffer retaining wall systems than soldier pile and lagging walls or sheet pile walls. Most secant pile walls in this project were designed to be cantilevered while multiple rows of tieback anchors would be required to control lateral wall deflection, if soldier pile and lagging walls or sheet pile walls were selected. The elimination of anchors on most cut walls was a major advantage to the project because of limited right-of-way and extensive utilities.

Fill Walls

The fill walls evaluated for this project included MSE and cast-in-place walls. MSE walls were the predominant type of fill walls used because of their lower cost and higher construction rate. Typically, granular soils were placed behind the MSE walls; however, in cases where significant fill will be placed on top of soft, highly compressible, organic soil deposits, lightweight foamed concrete fill (LFCF) will be used. A LFCF dry unit weight ranging between approximately 4.7 and 6.6 kilonewtons per cubic meter (kN/m^3) [30 and 42 pounds-per-cubic-foot (pcf) with an overexcavation of up to 1.8 m (6 ft) for load balancing resulted in a minimal net applied pressure to the bearing soil.

DEVELOPMENT OF SITE-SPECIFIC P-Y CURVES AND ANCHOR BOND STRENGTH FROM PRESSUREMETER TESTING

P-y Curves

The method for developing p-y curves from PMT data presented by Robertson et al. (1985) was used to derive the p-y curves from the corrected PMT data. This method incorporated the reloading segment of the corrected PMT data to calculate the soil resistance, P, and corresponding deflection, Y, using Equations (1) and (2) below. The Reduction Factor (α) was determined based on the soil type and depth where PMT was performed.

$$P = (\text{Corrected Pressure from PMT}) \times (\text{Pile Diameter}) \times (\text{Reduction Factor}, \alpha) \quad \text{.......(1)}$$

$$Y = \frac{\text{Corrected Volume from PMT}}{(2 \times \text{Initial Volume}) \times (0.5 \times \text{Pile Diameter})} \quad \text{.......(2)}$$

Undrained Shear Strength and Anchor Bond Strength

Baguelin et al. (1978) presented a nonlinear relationship between the undrained shear strength (s_u) and the net limit pressure (p^*_L), in Equation (3). The p^*_L, by definition, is the difference between the limit pressure (p_L) obtained from the PMT curve and the estimated horizontal total stress at rest (σ_{0H}), presented in Equation (4).

$$s_u = 0.67 \times \left(p_L^*\right)^{0.75} \quad \text{.......(3)}$$

$$p^*_L = p_L - \sigma_{0H} \quad \text{.......(4)}$$

where s_u and p^*_L are both in kilopascals (kPa)

The theoretical limit pressure is defined as the pressure reached for an infinite expansion of the PMT probe. The limit pressure is practically interpreted as the pressure reached when the soil cavity has been expanded to twice its initial size in accordance with Federal Highway Administration (FHWA, 1989) guidelines. Interpolation or extrapolation using the hyperbolic curve fitting method was performed to obtain this pressure. The horizontal total stress was also interpreted from the PMT curve.

The ultimate bond strength between the clayey soils and tieback anchors can be estimated from the in-situ undrained shear strength based on either FHWA (FHWA, 1999) or Post-Tensioning Institute (PTI, 1996) guidelines. The anchor bond strength depends largely on the installation methodology of anchors (i.e. gravity-grouted or pressure-grouted). According to the project specifications, all tieback anchors are required to be pressure-grouted. As a result, all bond strength values presented in this paper are relevant to pressure-grouted tieback anchors.

The ultimate bond strengths estimated from the in-situ undrained shear strengths of clayey soils are summarized in Table 1. A generalized bi-linear relationship presented in Figure 1 was used to estimate the ultimate anchor bond strength. As shown in the figure, the anchor bond strength increases with increasing undrained shear strength up to a limiting value. Undrained shear strengths from in-situ testing were typically greater than undrained shear strengths derived from conventional testing. Higher undrained shear strengths resulted in a reduced anchor bonded length, less right-of-way acquisition, and a cost savings for the project.

The preliminary results from the anchor load tests performed as a part of the Clybourn construction contract indicate that the estimated bond length from the in-situ testing was about 10% greater than the actual bond length required from the load tests.

TABLE 1. Summary of Pressuremeter Tests Performed on Clayey Soils for the North Leg, West Leg, and Core Segments of the Marquette Interchange Project

Clay Consistency	Number of Tests in Soil Type	s_u [a] (kPa)			s_a [b] (kPa)		
		Min	Max	Average	Min	Max	Average
(1)	(2)	(3)	(4)	(5)	(6)	(7)	(8)
Medium Stiff	9	72	144	107	103	207	154
Stiff	21	72	173	118	103	248	170
Very Stiff to Hard	27	77	392	229	110	276	261

Notes: [a] s_u = Undrained Shear Strength; [b] s_a = Anchor Bond Strength

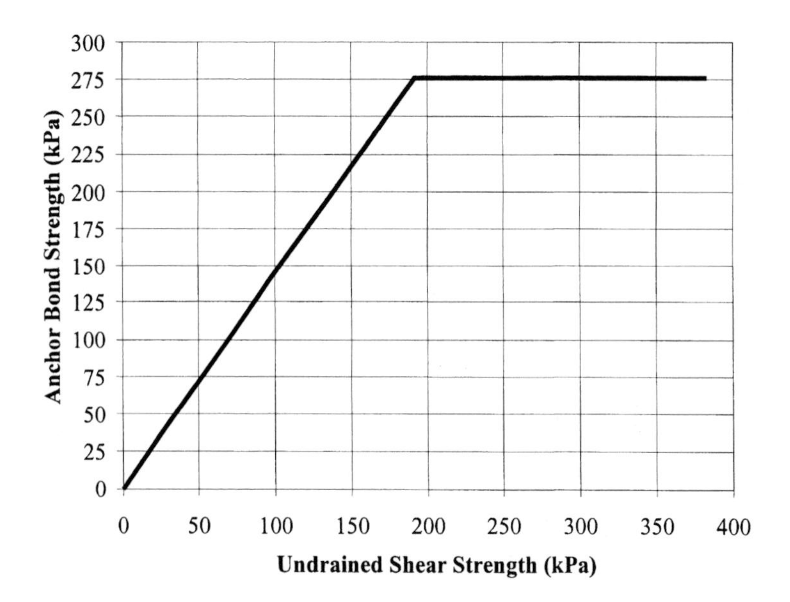

FIG. 1. Generalized Bi-linear Relationship between the Ultimate Anchor Bond Strength and Undrained Shear Strength of Clayey Soils in the Marquette Interchange Project

DEVELOPMENT OF FAMILIES OF PROJECT-SPECIFIC P-Y CURVES

Three families of project-specific p-y curves was developed from all the data obtained in the Marquette Interchange Project. The primary purpose of developing this family of p-y curves was to generalize the soil resistance derived from all PMT data in similar soil types for use in designing cut walls and pile-supported walls on this project. Due to the difference in subsurface characteristics and inconsistency in field PMT procedures, data obtained from the North Leg segment of the project and from the West Leg and Core segments of the project were developed into two families of p-y curves for retaining walls. Nine PMT were also conducted for the analyses of deep foundations supporting bridge structures in the Core segment of the project. These tests were used to develop into a third family of p-y curves. Among these three families, the p-y curves in the West Leg and Core retaining walls are presented hereafter.

A total of 12 categorized p-y curves were developed for the West Leg and Core segments based on the soil types and consistency of clay or relative density of sand. These curves, summarized in Table 2, were recommended for use in the design of cut walls in these segments. Each categorized p-y curve was derived by hyperbolic curve fitting of one or multiple site-specific p-y curve(s) from the pressuremeter testing. Graphical presentation of some of these categorized p-y curves and their site-specific data are presented in Figure 2. Because the curves are dependent on the size of the wall elements, the p-y curves presented in this paper are only for 36-inch diameter drilled shafts. It is also noted that a p-multiplier of 0.5 should be used along with these p-y curves to design secant pile walls because of the group effect of closely spaced side-by-side piles under lateral loading.

TABLE 2. Recommended Categorized P-y Curves to Design Tangent/Secant Drilled Shaft Walls in the West Leg and Core Segments of the Marquette Interchange

P-y Curve (1)	Soil Type (2)	C / RD[a] (3)	N[b] (blows/0.3m) (4)	s_u[c] (kPa) (5)
PMTPY WC-1	Sandy Silt Fill	Stiff	31[d] (9~15)	86[d] (48-96)
PMTPY WC-2	Silty Clay Fill	Very Stiff	6 (16-30)	182 (96-191)
PMTPY WC-3	Silty Clay Fill	Hard	49 (>30)	393 (>191)
PMTPY WC-4	Organic Silty Clay	Stiff	9-23 (9-15)	72-77 (48-96)
PMTPY WC-5	Silty Sand	Medium Dense	26 (11-30)	N/A
PMTPY WC-6	Silt and Sand	Very Dense	N/A (>50)	N/A
PMTPY WC-7	Silty Clay	Stiff to Very Stiff	9 (9-30)	81-100 (48-191)
PMTPY WC-8	Silty Clay	Stiff to Very Stiff	21 (9~30)	153-187 (48-191)
PMTPY WC-9	Silty Clay /Clayey Silt	Very Stiff to Hard	17-40 (>16)	177-191 (>96)
PMTPY WC-10	Silty Clay /Clayey Silt	Very Stiff to Hard	17-22 (>16)	220-230 (>96)
PMTPY WC-11	Silty Clay	Hard	32-34 (>30)	287-345 (>191)
PMTPY WC-12	Silty Clay	Hard	44 (>30)	325-373 (>191)

Notes: N/A = Test not applicable
[a] C / RD = Consistency or Relative Density of the Soil
[b] N = Standard Penetration N-value or Blow Count
[c] The first values in Column (5) are average site-specific data from testing. Values within the parentheses are common ranges based on consistency and relative density, according to Terzaghi, Peck, and Mesri (1996).

BENEFICIAL USE OF SITE-SPECIFIC P-Y CURVES IN DESIGN

Comparison between Conventional and Site-specific PMT P-y Curves

Conventionally, p-y models have been developed based on soil characteristics such as: type, consistency or relative density, shear strength parameters (c and/or ϕ), horizontal modulus of subgrade reaction (k_h), and shear strain at 50% of the maximum shear stress (ε_{50}). These models have been developed over the past 50 years and were discussed extensively in Reese and Van Impe (Reese, 2001). P-y curves based on the conventional models were compared with the site-specific PMT p-y curves at two retaining wall locations: R-40-301 and R-40-308. The comparison for R-40-308 is presented in this paper.

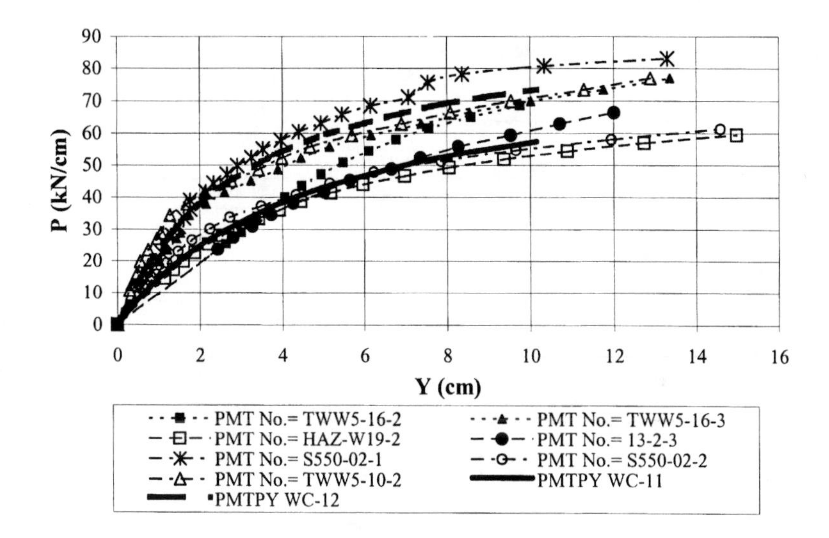

FIG. 2. A Family of Project-specific P-y Curves for Retaining Wall Design in the West Leg and Core Segment of the Marquette Interchange (PMTPY WC-11 to WC-12)

Subsurface profiles, soil types, and soil properties recommended in the MTP geotechnical reports for these two retaining walls (MTP, 2002) were used to generate the conventional p-y curves using the internal p-y curve module in the computer program, LPILE (Reese et al., 2000). At the testing location for wall R-40-308, the PMT was performed at elevations of 204, 200 and 195 m, Mean Sea Level (670, 655, and 640 ft, Mean Sea Level). The soil types at these three elevations were "Soil Type 1 above GWT", "Soil Type 2 above GWT", and "Soil Type 2 below GWT", respectively. The properties of these soil types as summarized in Table 3 were used to generate the conventional p-y curves at the same elevations where PMT p-y curves were developed. The comparisons between conventional p-y curves generated from LPILE and PMT p-y curves at wall R-40-308 are presented in Figure 3.

It can be observed in Figure 3 that the conventional p-y curves show higher resistance than PMT p-y curves at very small deflection values (Y< 7.6 mm or 0.3 inch). It is possibly contributed by factors such as testing inaccuracy, borehole imperfections, or oversized boreholes, the results of which are amplified at small deflections. These factors were corrected in subsequent PMT performed in the West Leg and Core segments of the project by a different testing subcontractor.

TABLE 3. Soil Types and Properties Used to Generate Traditional P-y Curves at Wall R-40-308

Soil Type (1)	Effective Unit Weight (kN/m^3) (2)	P-y Modulus k_{eq} (MN/m^3) (3)	Cohesion c_{eq} (kPa) (4)	Soil Strain $\varepsilon 50$ (5)
Type 1 above GWT	21.4	271	101	0.005
Type 2 above GWT	22.3	461	163	0.004
Type 2 below GWT	13.0	461	163	0.004

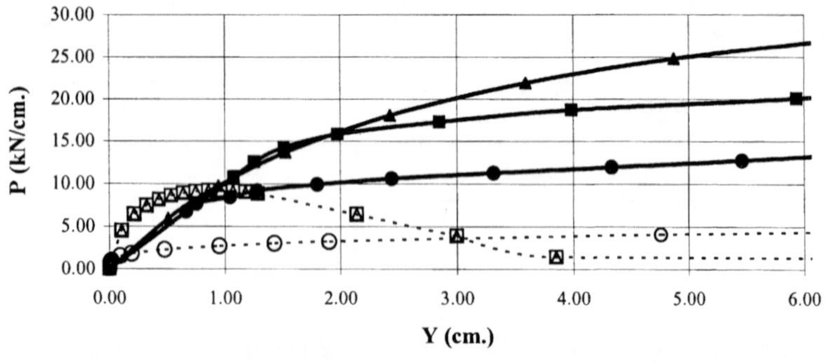

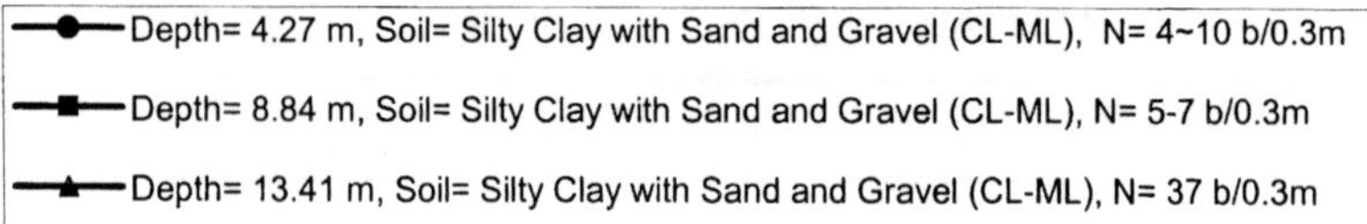

FIG. 3. Comparison Between Conventional and Site-Specific PMT P-y Curves at Wall R-40-308

As the deflection increases beyond the threshold value of 7.6 mm (0.3 inch), the PMT p-y curves present much higher resistance than conventional p-y curves. For a typical wall height between 6.1 and 9.1 m (20 and 30 ft) in the North Leg segment of the project, the limiting deflection at the top of wall is between 51 and 102 mm (2 and 4 inches) or 1% of the wall height. The deflection at the excavation level in front of the wall is expected to range between 12.7 and 25.4 mm (0.5 and 1 inch). Therefore, higher soil resistance, from PMT p-y curves, corresponding to a deflection greater than the threshold [7.6 mm (0.3 inch)] will be used in analyses and test-related factors stated above will have minimal effects on wall design. Generally, higher soil resistance from PMT p-y curves with deflection greater than the threshold value will provide a more economical design, as demonstrated hereafter.

Case Study

A case study was performed to demonstrate the cost effectiveness of using PMT p-y curves in retaining wall design. A typical wall section with a cut height of 7.9 m (26 ft) was used for this demonstration. The subsurface profile in the vicinity of walls R-40-301 to R-40-304 was used to analyze this cut wall. This subsurface profile consists of Soil Type 1 (Medium Stiff to Stiff Silty Clay) overlying Soil Type 2 (Stiff to Very Stiff Silty Clay). Groundwater table was assumed to be at a depth of 1.5 m (5 ft) below the top of wall.

This wall section was analyzed with both conventional p-y curves generated internally by the computer program, LPILE, using soil parameters recommended for these two soil types and site-specific PMT p-y curves. Using the same embedment depth of 80% of the exposed height, the maximum deflections at the top of wall from the computer program using conventional p-y curves and PMT p-y curves were 102 and 76 mm (4 and 3 inches), respectively. For this particular wall section, a deflection value of 102 mm was greater than the limiting deflection of 1% of the exposed height or 79 mm (3.1 inches). In order to lower the wall deflection to a value less than the limiting value, tieback anchors or an increased embedment depth would typically be required. On the other hand, analyses performed using PMT p-y curves resulted in a deflection value less than the limiting value; hence, eliminating the need for tieback anchors.

Furthermore, if the same deflection criterion was fulfilled in design, using PMT p-y curves to realistically model soil resistance, can potentially reduce the wall embedment and moment capacity of the wall element up to 30%. The reduction in wall embedment and moment capacity resulted in cost saving in wall design. The percentage of saving is expected to vary for different walls with varying wall heights, subsurface profiles and soil parameters, surcharge loads, and other factors.

DIRECT SETTLEMENT ANALYSES OF FILL WALLS USING CPTU SOUNDING DATA

Settlement of a few fill walls bearing on soft ground in this project was estimated using CPTU sounding data based on the method presented by Sanglerat (1972) and Costet and Sanglerat (1975). This method incorporates the measured cone tip resistance (q_c) and generated pore pressure (u_2) to estimate the constrained moduli (M) of soils. Different soil compressibility coefficients (α) as presented in Table 4 are assigned to different types of soils in estimating the constrained moduli, which also account for the long-term consolidation settlement of clays and organic soils.

TABLE 4. Soil Compressibility Coefficients

USCS Soil Type (1)	CL (2)	CL (3)	CL (4)	ML (5)	ML (6)	CH (7)	OL (8)	SM (9)	SM (10)
q_T^a (bars)	<7	7-20	>20	<20	>20	-		<50	>100
α^b	5.5	3.5	1.75	2	45	4	5	2	1.5

[a] q_T = Corrected cone tip resistance; [b] α = Soil Compressibility Coefficient

An example spreadsheet was used to estimate the potential settlement of walls R-40-338/339, which are back-to-back MSE walls that include lightweight foamed concrete fill, based on CPTU sounding data is presented in Table 5. The example assumes that the maximum height of these back-to-back MSE walls is 6.7 m (22 ft); the width of the fill is 13.1 m (43 ft), and an overexcavation up to 1.8 m (6 ft) for load balancing is planned to minimize the net pressure applied on the soft ground. The simplified 2 to 1 stress distribution method, presented in Holtz and Kovacs (1981) was used to estimate the stress increase under the loaded area. An estimated settlement of 56 mm (2.2 inches), with ranges between 25 and 102 mm (1 and 4 inches) was concluded from this analysis. This range of settlement is less than what was estimated from conventional analysis using parameters derived from laboratory consolidation tests. The results of using in-situ testing in retaining wall design in this particular example is the elimination of ground improvement measures or deep foundations to support the MSE walls and fill, which presents great cost saving to the project.

TABLE 5. Abbreviated Spreadsheet to Estimate Settlement of MSE Walls Based on CPTU Sounding Data at CP-2E-01

Depth (m) (1)	q_c [a] (bars) (2)	u_T [b] (bars) (3)	q_T (bars) (4)	Soil Type (5)	α_{CL} (6)	α (7)	M (bars) (8)	Δp [c] (bars) (9)	δ_{layer} [d] (mm) (10)
2.1	57.9	-0.2	57.9	SM		1.58	91.4	0.2	0.35
2.3	21.5	0.0	21.5	ML		4.50	97.0	0.2	0.33
2.7	7.4	0.1	7.4	CL	3.50	3.50	25.9	0.2	1.19
3.5	8.0	0.3	8.1	ML		2.00	16.2	0.2	1.80
3.7	4.2	0.0	4.2	CH		4.00	16.9	0.2	1.71
12.0	5.9	2.7	6.5	CL	5.50	5.50	35.6	0.1	0.52
13.0	28.1	0.5	28.2	SM		2.00	56.3	0.1	0.32
								Total	56.4

Notes: Data presented in the table is an abbreviated version and does not include all incremental depths. The complete spreadsheet includes data between a depth of 0.3 m (1 ft) and 13.1 m (43 ft), at an increment of 0.15 m (0.5 ft).

[a] q_c = Measured cone tip resistance

[b] u_T = Generated pore pressure measured behind the cone tip

[c] Δp = Net stress increase at depths

[d] δ_{layer} = Compression of sub-layers at depths

CONCLUSIONS AND REMARKS

Field instrumentation has and will continue to be implemented during construction. The instrumentation and monitoring data will verify the in-situ testing approaches used to optimize the retaining wall design presented herein. Nevertheless, the following conclusions can be drawn from the discussions in this paper.

1. In general, the PMT p-y curves in this paper present higher and more realistic soil resistance than conventional p-y models with input soil parameters from laboratory testing at a deflection greater than 7.6 mm (0.3 inch). The use of PMT p-y curves reduced the design wall embedment and moment capacity of the wall elements by 20 to 30% and eliminated or reduced the number of tieback anchors. The cost effectiveness of using in-situ testing in designing cut wall was well demonstrated in this paper.

2. The bond strength of tieback anchors, pressure-grouted in clayey soils, can be estimated from the in-situ undrained shear strength obtained from PMT. A generalized bi-linear relationship between anchor bond strength and undrained shear strength of clayey soils is presented in this paper. This relationship was used in estimating the bond strength and minimum bond length of tieback anchor in this project.

3. The preliminary results from the anchor load tests performed as a part of the Clybourn Advanced Construction Contract indicated that the estimated bond length from the in-situ testing was about 10% greater than the actual bond length required from the load tests.

4. Settlement analyses of MSE walls based on higher soil stiffness derived from CPTU sounding data resulted in a reduced magnitude of the potential settlement than conventional analyses with laboratory consolidation testing. These analyses eliminated the need for deep foundations to support the MSE walls, with the use of LFCF in the MSE walls.

5. The use of in-situ testing was a cost effective way of optimizing the retaining wall design. Roughly for every $1 spent on in-situ testing the estimated construction cost savings are between $10 and $35.

REFERENCES

Baguelin, F., Jezequel, J. F., and Shields, D. H. (1978). *The Pressuremeter and Foundation Engineering*, Clausthal-Zellerfeld, Trans Tech Publications, W. Germany.

Costet, J. and Sanglerat, G. (1975). "Cours Pratique de Mechanique des Sols. Tome 1: Plasticite et Calcul des Tassements." Dunod.

Feinstein, D. T., Eaton, T. T., Hart, D. J., Krohelski, J. T., and Bradbury, K. R. (2004). "Simulation of regional ground water flow in southeastern Wisconsin." Wisconsin Geological and Natural History Survey, Bulletin (number not yet assigned).

FHWA. (1989). *The Pressuremeter Test for Highway Applications*, Federal Highway Administration, McLean, Virginia, Publication No. FHWA-IP-89-008.

FHWA. (1999). *Geotechnical Engineering Circular No. 4 - Ground Anchors and Anchored Systems*, Federal Highway Administration, Publication No. FHWA-IF-99-015.

Holtz, R. D. and W. D. Kovacs. (1981). *An Introduction to Geotechnical Engineering*, Prentice Hall, Inc.

Mickelson, D M., Clayton L., Baker, R. W., Mode, W. N., and. Schneider, A. F. *Pleistocene Stratigraphic Units of Wisconsin*, University of Wisconsin Extension, Wisconsin Geologic and Natural History Survey, Miscellaneous Paper 84-1, 1-9, A7-1 to A8-3, and A10-1 to A10-4.

Milwaukee Transportation Partners. (2002). *Geotechnical Reports, Retaining Walls 301, 302, 303 and 308,* Milwaukee Transportation Partners for Wisconsin Department of Transportation.

Post-Tensioning Institute. (1996). *Recommendations for Prestressed Rock and Soil Anchors*, Post-Tensioning Institute.

Reese, L. C. and Van Impe, W. F. (2001). *Single Piles and Pile Groups Under Lateral Loading*, A. A. Balkema, Rotterdam.

Reese, L.C., Wang, S. T., Isenhower, W. M., Arrellaga, J. A., Hendrix, J. (2000). *Computer Program LPILE Plus (Version 4.0) – A Program for the Analysis of Piles and Drilled Shafts Under Lateral Loads: User's Guide and Technical Manual*, Ensoft Inc., Austin, Texas.

Robertson, P. K., Campanella, R. G., Brown, P. T., Grof, I., and Hughes, J. M. (1985). "Design of Axially and Laterally Loaded Piles Using In Situ Tests: A Case History." *Canadian Geotechnical Journal*, Vol. 22, (4), 518-527.

Sanglerat, G. (1972). *The Penetrometer and Soil Exploration*, Elsevier Publishing Co., Amsterdam.

Terzaghi, K., Peck, R. B., and Mesri, G. (1996). *Soil Mechanics in Engineering Practice*, John Wiley & Sons, Inc.

Williams, D. E. (1954). "Foundation Conditions in Downtown Milwaukee." PhD thesis, University of Wisconsin, 136.

SOIL NAILING: A LOCAL PERSPECTIVE

By Eric W. Bahner, P.E.[1], Member, Geo-Institute of ASCE

ABSTRACT: Since its introduction in Wisconsin in the late 1980's, soil nailing has become widely recognized as a cost-effective means of temporary and permanent earth retention. This paper discusses one geotechnical engineer's design philosophy, and describes important design and construction issues that have evolved over the past 15 years. A case history involving the design and construction of a 50 ft deep soil nail wall on the University of Wisconsin-Milwaukee (UWM) campus is also presented.

INTRODUCTION/HISTORY

The first reported use of soil nailing in the United States took place in Portland, Oregon in 1976, and gained popularity on both the east and west coasts through the mid-1980's. Soil nailing was introduced in Wisconsin on the Milwaukee Deep Tunnel Project between 1988 and 1990, where it was used to provide temporary support of 72 and 86 ft deep excavations for the construction of the NS-11 and NS-6 drop shafts. Soil nailing was introduced as a temporary earth retention method in Madison between 1990 and 1993, during the construction of a new State Administration Building, Grainger Hall on the University of Wisconsin - Madison campus and the Dane County Jail. All three projects required temporary support of excavations of 35 to 40 ft deep in a dense urban environment.

Since then, soil nailing has been used as a means of temporary earth retention for excavations as deep as 55 ft on construction projects with favorable subsurface conditions. The method has also been used in a permanent application on several

---

[1] Chief Engineer, Edward E. Gillen Company, 218 West Becher Street, Milwaukee, Wisconsin USA, Email: eric.bahner@gillenco.com

building and civil works projects, and has been used as a lateral support component for foundation underpinning schemes on many projects in southeastern and south-central Wisconsin. The author has used this method on approximately 5 to 10 percent of earth retention projects performed by his company since 2000.

This paper reflects the authors' design philosophies, presents issues critical to the successful construction of soil nail walls, and presents a case history of a soil nailing project recently completed as a part of the Klotsche Center Addition currently under construction on the UW-Milwaukee Campus.

DESIGN PHILOSOPHY

Feasibility

Soil nailing is a ground reinforcement technique that uses passive metal inclusions - nails - to improve the overall strength of a soil mass, restrain displacements, and limit decompression during and after excavation. Like reinforced concrete, soils that are generally suitable to soil nailing have substantial compressive strength, but very little, if any, tensile strength. The nails add this tensile strength component to the reinforced soil mass.

The stand-up time of the ground is critical to the success of soil nailing. Suitable ground for soil nailing must have sufficient strength and stand-up time to allow installation of the nails and placement of the shotcrete. In general, sufficient stand-up time is one to two days for a temporary cut of 5 to 6 ft, or long enough to allow installation of the nails and placement of the shotcrete. In general, well graded, medium to extremely dense granular soils and very stiff to hard clays are suitable for soil nailing. Uniformly graded, clean granular soils and soft clays are typically not. The author has been involved with soil nailing projects and based on his experience he consider soil nailing a feasible earth retention option if the following statements are true:

Granular Soil:

- The minimum Standard Penetration Test value (N) is greater than 10.
- The fines content (P200) exceeds 15 to 20 %.
- The uniformity coefficient is greater than 2.

Cohesive Soil:

- The Clay is of low plasticity and has a USCS classification of CL.
- The unconfined compressive strength is greater than or equal to 1.5 tsf.

The potential for surface water runoff and/or groundwater seepage is also important consideration. For example, large, unpaved areas with little relief could be areas where surface water could pond and eventually seep into and soften the reinforced earth mass. However, large paved areas that slope toward a construction site can convey large volumes of surface runoff into the excavation. Mottling in clayey soils could be indicative of a fluctuating groundwater table, a condition that could be very problematic both during and after construction of the wall is complete. Each of these conditions warrants special attention. In certain situations, such conditions could warrant the selection of a different earth retention method.

In situations where excavation must proceed immediately adjacent to and below the footing grade of an adjacent structure, it is common to incorporate helical piers or micropiles into the design to carry foundation loads below the base grade of the new excavation. In situations where the excavation is further removed from the adjacent existing structure, but within the foundation zone of influence, the soil nail design must be sufficient to resist the soil load and the superimposed foundation surcharge load.

Internal Stability

The spacing, length and capacity of the individual nails govern the stability of a soil nail wall. Nicholson (1986) recommends a maximum nail spacing of no more than 3 to 6 ft. A similar spacing range is described by Byrne et al. (1996).

For walls with simple geometry, we use the earth pressure method described by Thompson and Miller (1990), Elias and Juran (1991) and Byrne, et. al (1996). This method provides a rational basis for the design of the nails using standard earth pressure diagrams. We use the computer program SNAILZ to analyze walls with more complex geometry. SNAILZ is a limit equilibrium program developed by the California Department of Transportation (CALTRANS), Division of New Technology, Material and Research Office of Geotechnical Engineering.

External Stability

A soil nail wall behaves similarly to a mechanically stabilized earth (MSE) wall. An MSE wall is constructed using reinforcing elements such as geogrids or geotextiles. A soil nail wall is constructed in the existing soil using the drilled in place nails. To be externally stable, the resulting "gravity wall" must be sufficiently wide enough to resist overturning, sliding and classical slope failure.

Generally speaking, soil nail walls with length to wall height (L/H) ratios of 0.7 to 0.8 are externally stable.

Facing Design

The facing design is empirical and typically consists of a 4 to 6-inch layer of 4,000-psi shotcrete reinforced with welded wire mesh. Guidance concerning mix design and shotcrete application is available through American Concrete Institute (ACI). The reinforcement at the nail locations can include a second layer of wire mesh, continuous wale bars at the nail levels or a grid of short reinforcing bars at each nail location. The photographs in Figure 1 provide two examples of the reinforcement in a temporary shotcrete facing. A standard 4-inch thick shotcrete layer is satisfactory for most temporary applications.

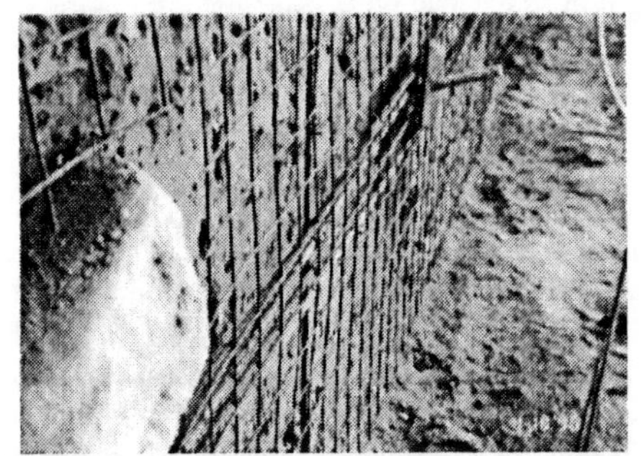

(a) Soil nail and reinforcement details for shotcrete

(b) Soil nail

FIG. 1. Shotcrete Reinforcing Details

Wall Performance

Elias and Juran (1991) report that facing displacements are similar to those measured for braced cuts in uniform ground conditions. This is generally consistent with our local experience in clayey soils. However, lateral movements can be greater in conditions where water seepage, and freezing and thawing occur. Conversely, in dense, self-supporting granular soils, very little movement is required to engage the nails; consequently, displacements are very small and sometimes not measurable.

CONSTRUCTION ISSUES

Very simply, soil nailing is a top-down construction method consisting of a repeating process of mass excavation in 5 ft lifts, installation of the specified nails, and finally, application of a reinforced shotcrete layer for that lift. The actual process can vary by contractor, and exposed ground conditions. The photographs of Figure 2 illustrate the basic installation process.

During construction, the stand-up time of the ground is critical. There are many different methods to install soil anchors in a variety of ground conditions; however, poor stand-up time has a dramatic impact on the overall construction process, schedule, quality of the finished product and ultimately, the cost. In cases where the ground with insufficient stand-up time is limited in extent or is sporadic in nature, construction can proceed by excavating and constructing the wall in smaller lifts and/or modifying the construction sequence. Ground improvement techniques, such as injecting grouting and jet grouting, have also been used to improve stand-up time.

Proper management of surface and subsurface water is critical to the successful construction and performance of any soil nail wall. Surface water runoff can have a catastrophic impact on the performance of a soil nail wall. Therefore, the construction team must devise a program to divert surface runoff away from the excavation, and keep it from saturating the areas immediately behind the excavation face. We have developed very specific requirements that must be addressed by our field crews, or by the General Contractor.

Subsurface water can be managed in a number of different ways including:

- Installing geocomposite drain board as the wall is constructed.
- Drilling weep holes through the shotcrete.
- Installing horizontal drains, such as slotted PVC and geotextile wicks.

(a) Soil excavation in 5 ft. lifts

(b) Installing the nails

FIG. 2. Soil nail wall construction process

(c) Placing reinforcement

(d) Applying the shotcrete

FIG. 2. (Cont.) Soil nail wall construction process

With the exception of the drain boards, drainage system can be installed as the wall is constructed, or after construction, if warranted. Permanent walls are typically designed with a drainage system consisting of drain board connected to weep holes through the wall, or plumbed to a nearby storm sewer manhole. Temporary walls are typically designed without drainage system. However drainage can be added, if warranted.

Assuming that the ground conditions are suitable, the best performing soil nail walls are typically those with a well thought-out plan to manage surface water runoff.

CASE HISTORY: UW-MILWAUKEE KLOTSHE CENTER ADDITION

General Information/Retention System Design

In the fall of 2003, UW-Milwaukee began work on a major addition to the north side of the current athletic facility, the Klotsche Center. The new addition required a 190 ft x 260 ft excavation extending to a depth of 45 ft below present grade, and to a depth of 35 ft below existing column footings on the north side of the existing structure.

The topography across the site was relatively flat; however, the areas along the north and west sides, and north half of the east side of the excavation were wooded and slightly lower in elevation than the building footprint. A site plan is provided in Figure 3.

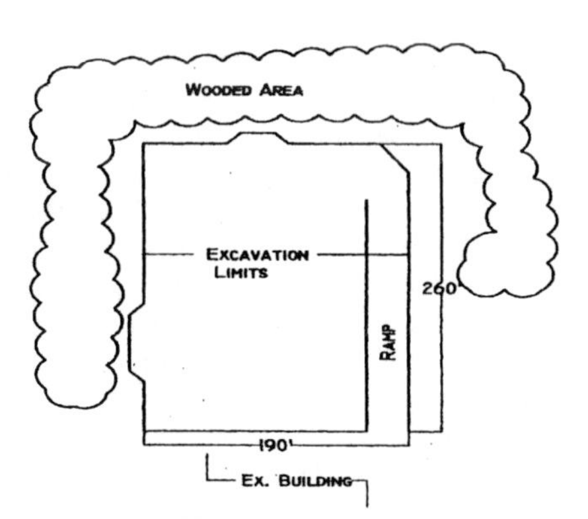

FIG. 3. Site plan

The project geotechnical report described a subsurface profile consisting of very stiff to hard, clay till with pockets/layers of sandy outwash deposits. The fines content of these deposits varied considerably, and were expected to be sources of seepage into the excavation. The specification required the retention system designer to account for the presence of such deposits in the design.

We evaluated several traditional earth retention systems for this project including driven sheeting and soldier piles and lagging, as well as soil nailing. Based on our experience in similar ground conditions, we believed that soil nailing was well suited to the anticipated ground conditions, and the most cost effective. In most applications, soil nailing is typically 25 to 30 percent less expensive than a tied back soldier pile and lagging wall.

The soil nail design for this project required approximately 900 nails with lengths ranging from 20 to 40 ft. The nails consisted of #7 to #10 thread bars in 8-inch diameter grouted holes. The facing consisted of an empirically designed, 4-inch thick reinforced shotcrete facing. The nails were designed to resist the lateral soil pressures for the full depth of the excavation plus a uniform surcharge load of 300 psf. This uniform surcharge load typically accounts for most temporary construction load conditions, including dump trucks and ready-mix concrete trucks. It does not account for large point loads or area loads such as those generated by backhoes and cranes. The nails for the south wall of the excavation were designed to resist both the lateral soil pressure for the excavation plus the lateral surcharge pressure resulting from the adjacent isolated column footings

Crane Surcharge Loads

Well after construction was underway, we learned that the general contractor wished to operate two Liebherr 1250 crawler cranes along the excavation perimeter in lieu of installing a tower crane for the site. For the anticipated pick loads, calculations using software provided by the crane manufacturer estimated that the resulting contact stresses on the tracks of the cranes would be approximately 9,000 psf, well in excess of the 300 psf uniform surcharge included in the design. We were very concerned that, without special precautions, operating such large cranes in close proximity to the earth retention system could precipitate unwanted lateral movement or catastrophic failure of the wall. Using information provided by the GC, we completed a multi-layer elastic analysis of the crane tracks, underlying soil and potential materials to reduce the magnitude of the contact stresses and developed operating constraints for the cranes when operating on and off timber crane mats. We also required that any pivoting of the crane on the tracks be undertaken at distances no closer than 20 ft of the excavation to avoid dynamically loading the retention system in a manner that we could not quantify. These operating constraints were presented to the GC's project team and discussed with the crane operators. The behavior of the excavation was monitored by optical survey on a daily basis. The GC, their crane operators and our crew, did more informal monitoring of the wall visually.

Construction Challenges

Construction of the retention system began in the fall of 2003 without incident. However, as work continued to excavation base grade, outwash deposits with poor stand-up time limited our progress. Although current drilling technology will allow the installation of anchors in a variety of ground conditions, ground with poor stand-up can severely hamper construction of the shotcrete facing, and if the condition is wide spread, force the selection of a completely different retention method. When necessary, we improved the stand-up time of these deposits by drilling closely spaced, vertical spiles. The spiles in this application are essentially vertical nails. The spiles improve the stand-up time of soil by forcing arching between these elements.

As we completed our work, and later during basement construction, repeated freezing and thawing and related soil softening resulted in lateral wall movements of approximately 1-1/2 inches on the north half of the excavation. However, unexpected amount of seepage—estimated to be as much as 5,000 to 7,000 gallons/day—resulted in wall movements as large as 12 inches in an isolated portion of the northeast excavation corner. By contrast, the wall movements measured on the south half of the excavation were typically 1/2 inch or less.

In the northeast corner of the excavation, we employed several methods to drain the water from the area, including weep holes drilled through the shotcrete, and horizontal wick drains. Although these attempts helped drain the soil to some extent, the sheer volume of water overwhelmed these systems. Fortunately, this area somewhat removed from the building footprint, and did not interfere with the construction. The area was bermed with crushed stone to arrest further lateral movement.

Aside from seepage-related wall movements, the retention system performed satisfactorily, and the GC completed basement construction in the spring of 2004.

CONCLUSIONS

Since its introduction in Wisconsin in the early 1990's, soil nailing has proven to be a reliable and cost-effective retention system for both temporary and permanent applications for excavations as deep as 86 ft. Soil nailing, though very cost effective, is not feasible in all ground conditions. Based on our experience, we offer the following "lessons learned":

- The ground conditions must be suitable for soil nailing. If "most" of the ground is suitable, but some areas are not, careful consideration is required before choosing this approach. Let the ground conditions dictate the method of support.
- Stand-up time is critical to the successful construction of soil nail walls. Improper evaluation of stand-up time can result in construction delays, excessive construction costs and even abandonment of the retention method. Meaningful subsurface information is essential in making such evaluations.
- Surface water must be effectively controlled. Back slopes behind the excavation must be protected to prevent seepage behind the shotcrete face. Be careful of large, open areas where water can pond behind an excavation. Ponded water can soften the reinforced soil mass that makes up a soil nail wall, and increase the magnitude of wall movement.
- Subsurface water, if present, must be allowed to drain through the facing. In most temporary applications, drainage is not necessary. However, if seepage is observed, appropriate measures must be taken.
- The measured lateral movements of soil nail walls in dense granular soils have typically been very small. Lateral movements in cohesive soils have proved consistent with those associated with more conventional retention systems. Surface water runoff, and surface and subsurface seepage, especially in concentrated areas can greatly exacerbate wall movement. The performance of soil nail wall for the Klotsche Center addition provides a dramatic illustration.
- It is not economical to design soil nail walls for large surcharge loads. However, it is possible to devise ways to reduce heavy surcharge loads to acceptable levels.
- Soil nailing is a craftsmanship-intensive earth retention method. The skill and expertise of the construction crew is just as important as the details of the wall design. Experience is critical.

ACKNOWLEDGEMENTS

The author acknowledges the University of Wisconsin-Milwaukee and the State of Wisconsin, Department of Facilities Development, the owner and developer of the project.

Miron Construction Company of Neenah, Wisconsin is the General Contractor. Edward E. Gillen Company was the successful earth retention system design/ build subcontractor for General Contractor (GC) Miron Construction Company of Neenah, Wisconsin. As with any successful earth retention project, teamwork is essential between the GC, earthwork contractor and retention system contractor.

Finally, the author acknowledges the expertise and conscientiousness of the Gillen Company crew.

REFERENCES

Bahner, Eric W. (1994). "Soil Nailing in Madison, Wisconsin". *Proceedings-- 1994 Technical Symposium at U.S.-Based Shareholders' Informational Meeting.* Woodward-Clyde Consultants.

Byrne, R.J., et. al (1996). Manual for Design and Construction Monitoring of Soil Nail Walls, FHWA Report FHWA-SA-96-069, November 1996.

Christensen, R.W. (1998 to 1993). Personal Communications.

Elias, Victor, and Juran, Ilan (1991) *Soil Nailing for Stabilization of Highway Slopes and Excavations*, FHWA Report RD-89-198.

Nicholson, Peter J. (1986). "In-situ Ground Reinforcement Techniques", presented at The International Conference on Deep Foundations, Beijing, China, September 1986 (Meeting Reprint).

Thompson, Steven R. and Miller, Ian R. (1990). "Design, Construction and Performance of a Soil Nailed Wall in Seattle, Washington." *Design and Performance of Earth Retaining Structures*, Special Publication No. 25, ASCE, pp. 629-643.

GEOSYNTHETIC REINFORCEMENT-COHESIVE SOIL INTERFACE DURING PULLOUT

Izzaldin Almohd[1], Murad Abu-Farsakh[2], and Khalid Farrag[3]

ABSTRACT: The influence of soil's physicochemical property (cohesion) on the pullout behavior of different geosynthetic reinforcements was studied in this paper. This was accomplished by considering the pullout load-displacement curves measured at different points along the reinforcement. One woven geotextile and four geogrids with different stiffnesses and geometries were studied. The measured load-displacement and the deduced load-deformation curves were examined to determine the interface strength parameters (interface adhesion and friction). The influence of the interface adhesion on the load-deformation curve for a given segment of the reinforcement was indicated by an inflection point that corresponds to a compatibility force. The compatibility force is the force required to produce displacements at both ends of the segment. The load-deformation curves were bilinear for relatively weaker reinforcements and nonlinear for stronger reinforcements. The compatibility forces were used to back-calculate the reinforcement-soil interface adhesion and the angles of interface friction. The angles of interface friction were found to be inversely proportional to the squared root of the confining (overburden) stress level of the tests.

INTRODUCTION

Understanding the mechanisms of interaction at the interface between the geosyntehtic reinforcement and the soil, and the contributions of each mechanism of interaction can lead to a more reliable and cost effective design. The internal stability of a geosynthetic reinforced soil mass should assure that the reinforcement will neither break nor pullout due to external and internal stresses. This requires considerations for the reinforcements' intensity, length, and strength at different stress levels. The interface or resistance depends on the soil type and strength, the reinforcement material and surface roughness, and the compatibility between the soil and the reinforcement.

The majority of the studies pertaining to the reinforcement-soil interface interactions have been conducted on granular (select) soils. This is mainly due to the drainage and strength characteristics of these materials. However, the cost of these backfill materials compared to marginal cohesive soils could be a major concern.

---

[1] Research Associate, Louisiana Transportation Research Center, Baton Rouge, LA.
[2] Research Assist. Prof., Louisiana Transportation Research Center, Louisiana State University, Baton Rouge, LA.
[3] Manager, Civil and Geotechnical Research, Gas Technology Institute, Des Plaines, IL.

Using marginal soils could significantly reduce the cost of construction of geosynthetic-reinforced soil (GRS) walls. This calls for more research to investigate the feasibility of using soils with considerable fines and cohesive properties as an alternative backfill material. In the state of Louisiana, a research was started in 1994 at the Louisiana Transportation Research Center (LTRC) aiming at investigating the potential of using marginal cohesive soils as reinforced backfills. Full-scale (field) and reduced-scale (laboratory) pullout tests were conducted on various geosynthetic reinforcements with different lengths of 3, 4, and 5 ft (0.9, 1.2 and 1.5 m, respectively) and under various normal stresses ranging from 3 psi (20.7 kPa) to 15 psi (103.4 kPa). The preliminary results were reported in Farrag and Morvant (2004) and Mohiuddin (2003), whereas the analyses of the tests were reported in Abu-Farsakh and Almohd (2004).

In this paper, the pullout load-displacement curves for different geosynthetic pullout strips will be presented and analyzed to demonstrate the influence of soil cohesion on interface response. This necessarily requires transformation of the load-displacement into load-deformation curves using the measured displacements of every two points along the reinforcement segments. Only the results of the laboratory pullout tests that were instrumented with displacements gages will be presented and analyzed in this paper.

Material properties, instrumentation, and Test Program

Properties of Backfill

The backfill soil is classified as low plasticity clay (AASHTO A-4), with a plasticity index (*PI*) equal to 15%, and a liquid limit (*LL*) of 42%. The soil has a maximum dry density of 105 pcf (16.5 kN/m^3) at an optimum moisture content of 18.5% per Standard Proctor. The results of direct shear tests, on samples prepared at the optimum density and moisture, revealed an angle of internal friction of 24°, and a cohesion intercept of 300 psf (14.4 kPa).

Properties of Reinforcements

Four types of geogrids (Stratagrid-500, UX-750, UX-1500, and UX-1700, in respective order of increasing strength and stiffness), and woven 4 x 4 geotextile were used for the study. The basic material properties for these reinforcements are provided in Farrag and Morvant (2004) and Mohiuddin (2003). The laboratory pullout specimens were 1 ft (30 cm) wide and 3 ft long, each.

Instrumentation and Test Schedule

Each of the pullout specimens was equipped with three linear variable displacement transducers (LVDT's); at the point of load application, at 1 ft (30 cm), and at 2 ft (60 cm) from the point of loading (Figure 1). The pullout strips were tested at different normal stresses ranging from 3 to 15 psi (20 to 103 kPa).

METHODOLOGY FOR EVALUATING INTERFACE PARAMETERS

The measured load-displacement curves at any two points along the reinforcement will be used to deduce the load-deformation curve for the reinforcement segment bounded between these two points. Referring to Figure 2, the differential displacement, δu_{ij}, measured between any two points (i and j) along the reinforcement corresponds to the material distortions (deformations or extensions), δu_{ij}. The presence of the interface adhesion (C_a) at the soil-reinforcement will results in partial strain compatibility between the soil and the reinforcement. This means that the reinforcement strains are the same as those of the soil at the immediate interface. The soil moves or deforms with the reinforcement, thus the resulting reinforcement-soil relative movement is negligible, and the resulting interface friction angle is close to zero. Once this value is exceeded, relative movements (slippage) between the soil and the reinforcement start to develop, leading to the mobilization of interface friction. Since reinforcement strains can be related to the reinforcement modulus, the compatibility strain will correspond to a compatibility force which can be defined as the force required to overcome the soil-reinforcement interface adhesion and will be referred to as the compatibility force, P_o. The equivalent stress (the force, P_o, per unit area of the reinforcement-soil interface between the two measurement points) will be approximately equal to the average interface adhesion (C_a). Figures 3a and 3b illustrate the influence of the interface adhesion on the shapes of the load-displacement and the load-deformation curves for a portion of the reinforcement strip.

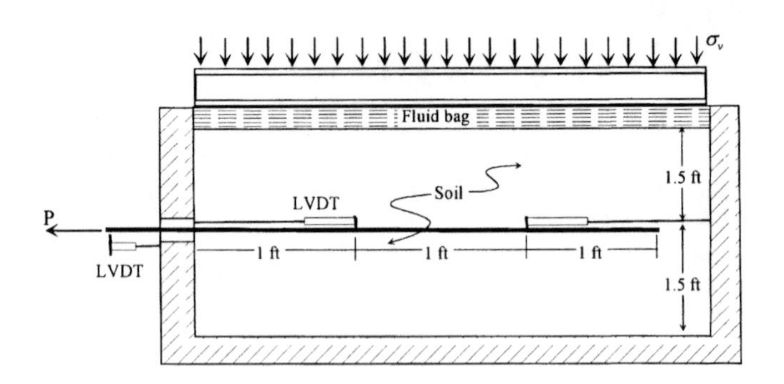

FIG. 1. A schematic of pullout test setup and the instrumentation plan. (After Farrage and Morvant, 2004)

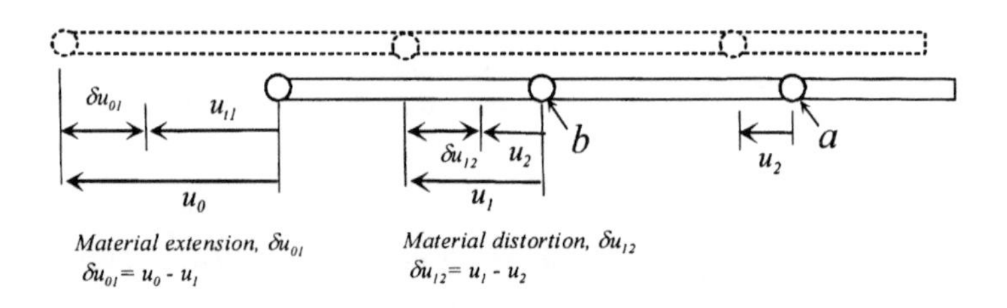

FIG. 2. Translational and distortional movements of reinforcement

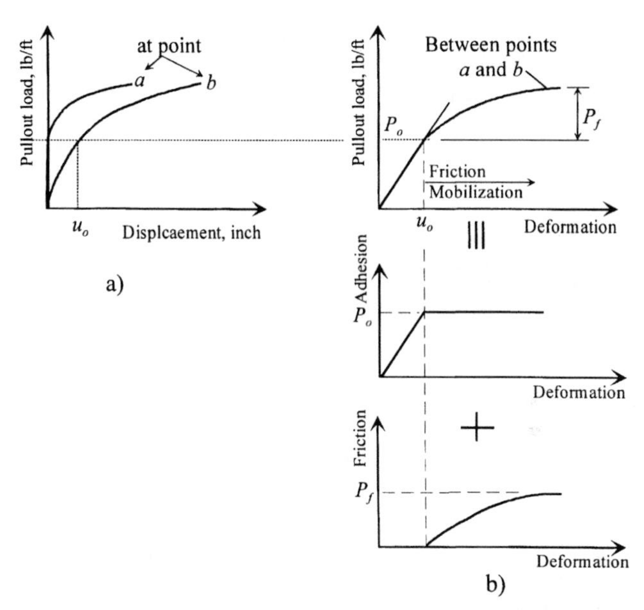

FIG. 3. Effects of soil's cohesion on the: a) load-displacement curve, and b) load-deformation curve and the separation of friction and adhesion resistances

ANALYSIS OF PULLOUT CURVES

Load deformation curves were produced from the measured load-displacement curves for all laboratory test specimens. An example demonstrating the development of load-deformation curves for the UX-750 geogrids is provided in Figure 4. In general, the load-deformation curves revealed that under low pullout loads, the far end of the reinforcement will be fixed with zero movement until the applied load reaches a certain value, which will be referred to as compatibility force, P_o. Physically, this force would be equivalent to the load required to overcome the adherence between the soil and the reinforcement for the reinforcement-soil contact area bounded between the two measurement points. It marks the evolution of the relative movements between the soil and reinforcement responsible for the mobilization of interface friction. The value of this load for a given type of soil and reinforcement depends primarily on the soil physicochemical property (cohesion) and the type and geometry of the reinforcement material, and to a less extent, the level of stress.

The approximate interface adhesion strength (C_a) for each type of reinforcement was calculated by dividing the compatibility force by the total interface area of the 2 ft (0.60 m) long reinforcement segment. The compatibility (adhesion) force and the deduced adhesion strength for each type of reinforcement are listed in Table 1 and also depicted in Figures 5a and 5b as a function of the normal stress level. The interface strength parameters are the apparent parameters calculated based on the total area without considering the areas of the apertures. To examine the goodness of this assumption, the remaining resistance, which is defined as the difference between the pullout resistance and the adhesion intercept ($\tau - C_a = (P_r - P_o)/L$), are represented as a function of the normal stresses in Figure 6. The remaining resistance is shown to be linearly dependent upon the overburden pressure as follow:

$$\tau - C_a = \begin{cases} 0.25\sigma_n^{'} \cdots\cdots\cdots \text{Woven geotextile} \\ 0.15\sigma_n^{'} + 3.3 \cdots\cdots \text{Stratagrid} - 500 \\ 0.33\sigma_n^{'} \cdots\cdots\cdots \text{UX} - 750 \\ 0.15\sigma_n^{'} + 1.3 \cdots\cdots \text{UX} - 1500 \\ 0.15\sigma_n^{'} + 3.3 \cdots\cdots \text{UX} - 1700 \end{cases} \quad (1)$$

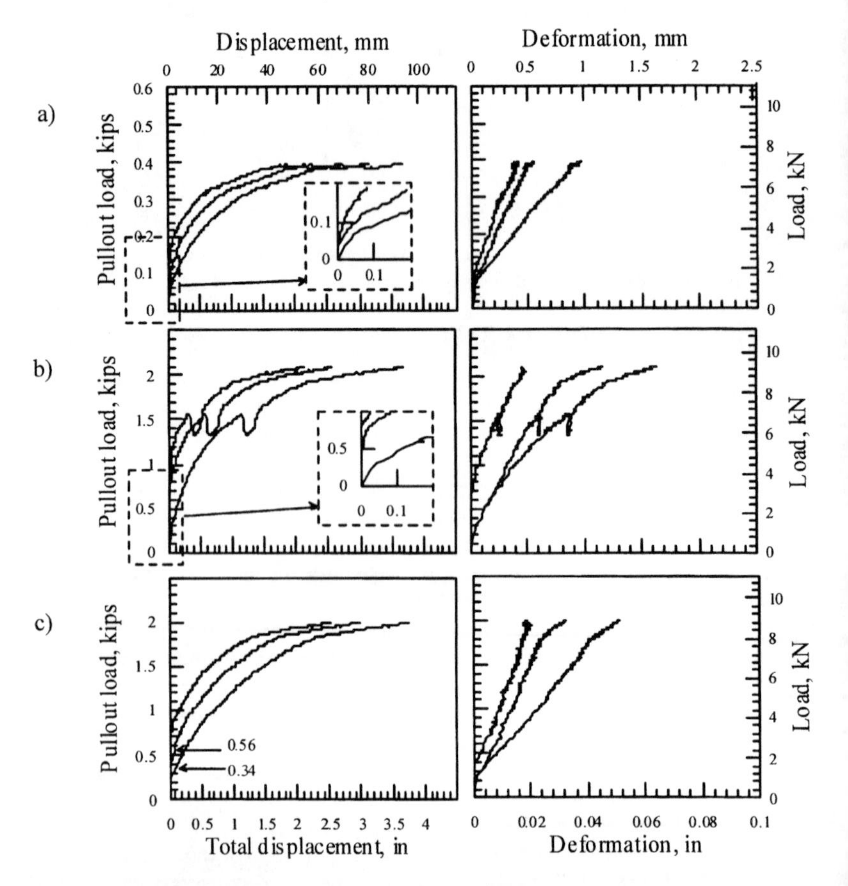

FIG. 4. Load displacement and average load deformation curves for UX 750 geogrid under normal pressure of: a) 3 psi, b) 5 psi, and c) 7 psi

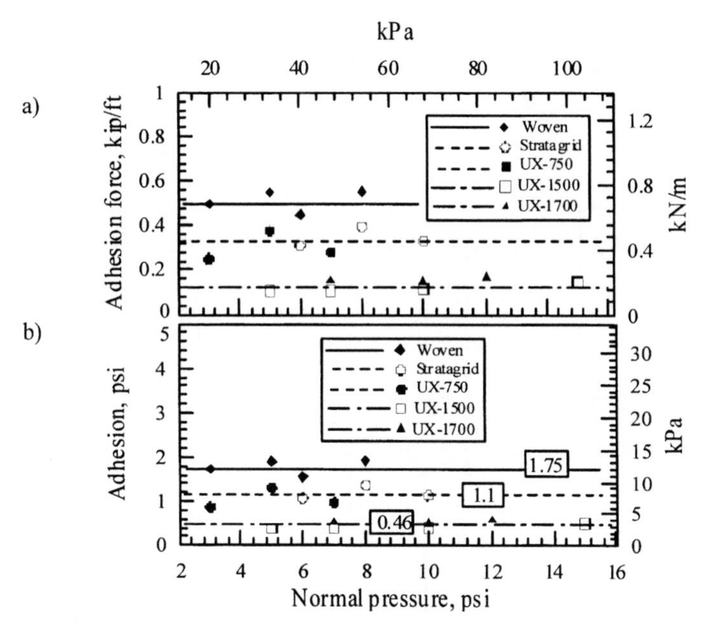

FIG. 5. Interface bonding versus normal stress levels based on laboratory load-deformation curves: a) adhesion force, b) adhesion strength (stress).

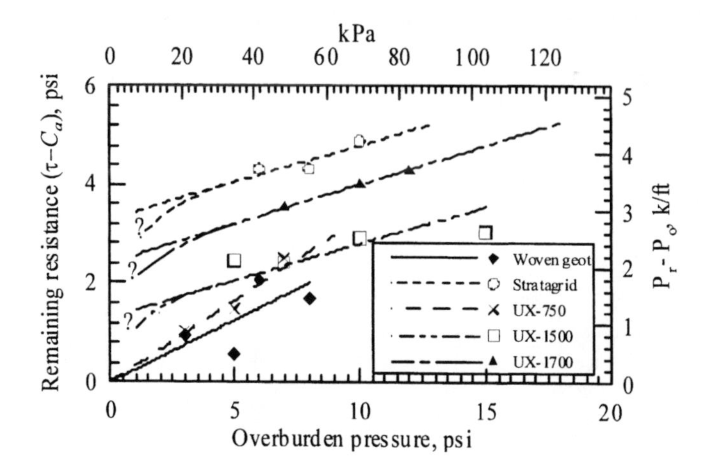

FIG. 6. Remaining stress as a function of overburden stresses based on load-deformation curves

TABLE 1. Reinforcement-soil interface parameters based on laboratory load-deformation curves

Reinf.	Overb. psi	P_r k/ft	P_o k/ft			Adhesion, C_a psi			Average		δ deg	C_a/C %	$\tan\delta/\tan\phi$ %
			0-1	0-2 *	1-2	0-1	0-2	1-2	P_o, k/ft	C_a, psi			
Woven 4X4	3	2.31	---	1	---	---	1.74	---	0.50	1.74	17.3	83.3	70.2
	5	2.15	---	1.1	---	---	1.91	---	0.55	1.91	6.6	91.7	26.0
	6	3.14	---	0.9	---	---	1.56	---	0.45	1.56	19.0	75.0	77.5
	8	3.14	---	1.11	---	---	1.93	---	0.56	1.93	12.0	92.5	47.9
Stratagrid 500	6	4.66	0.36	0.62	0.26	1.25	1.08	0.9	0.31	1.08	35.7	51.7	161.5
	8	4.92	0.35	0.79	0.44	1.22	1.37	1.53	0.40	1.37	28.4	65.8	121.3
	10	5.22	0.38	0.66	0.28	1.32	1.15	0.97	0.33	1.15	26.1	55.0	109.9
UX-750	3	1.63	0.31	0.5	0.19	1.08	0.87	0.66	0.25	0.87	18.7	41.7	76.2
	5	2.4	0.33	0.75	0.42	1.15	1.3	1.46	0.38	1.30	16.4	62.5	66.3
	7	3	0.34	0.56	0.22	1.18	0.97	0.76	0.28	0.97	19.6	46.7	80.2
UX-1500	5	2.43	0.13	0.21	0.08	0.45	0.36	0.28	0.11	0.36	26.1	17.5	109.9
	7	2.41	0.13	0.21	0.08	0.45	0.36	0.28	0.11	0.36	19.1	17.5	77.8
	10	2.86	0.1	0.23	0.13	0.35	0.4	0.45	0.12	0.40	16.2	19.2	65.4
	15	3.06	---	0.29	---	---	0.5		0.15	0.50	11.4	24.2	45.5
UX-1700	7	3.5	0.16	0.33	0.13	0.56	0.57	0.45	0.15	0.53	26.7	25.3	113.0
	10	3.89	0.15	0.29	0.14	0.52	0.5	0.49	0.15	0.50	21.8	24.2	89.8
	12	4.19	0.15	0.336	0.18	0.52	0.58	0.63	0.17	0.58	19.6	27.7	79.9

* For 2 ft long segment.

The coefficients (multipliers) on the right side of Eq. (1) are the coefficients of friction of the soil-reinforcement interface. They correspond to the angles of interface friction (δ) of 14, 8.5, 18.26, 8.5, and 8.5 for the woven geotextile, stratagrid-500, UX-750, UX-1500, and UX-1700 geogrids, respectively. These values are different from the values in Table 1. Accordingly, the assumption that the use of total areas may substitute for the bearing contributions of the transverse members of the reinforcements may not hold. The intercepts, on the other hand, for the stratgrid-500, UX-1500, and UX-1700 geogrids, are only based on curve fitting. These intercepts can be attributed to the bearing (passive) resistances of the transverse members of the geogrids. The bearing resistance of transverse members is a function of the soil's shear strength parameters, the thickness and stiffness of the geogrid reinforcement, number of transverse members, and, to less extent, the level of normal stress. The bearing resistances should be considered, and the total pullout resistance should be expressed as:

$$P_r = friction + adhesion + bearing$$
$$P_r = (\sigma_n^{'} \tan\delta + C_a)A + P_b \qquad (2)$$

P_b is the bearing contribution to the pullout resistance by the transverse members. It's a function of the soil's shear strength parameters and geometry of the transverse members (number, n, of members; width, w, of members; and thickness, d, of members) as follow (Peterson and Anderson, 1980):

$$\mathrm{P_b = nwd(CN_c + \sigma_n' N_q)} \qquad (3)$$

where N_c and N_q are functions of the angle of internal friction of the soil, given as:

$$\mathrm{N_q} = \begin{cases} e^{\pi \tan\phi} \cdot \tan^2(45 + \frac{\phi}{2}) \text{.....bearing failure mode (Peterson and Anderson, 1980)} \\ e^{(\frac{\pi}{2}+\phi)\tan\phi} \tan^2(45 + \frac{\phi}{2}) \text{..punching failure mode (Jewell et al., 1984)} \end{cases} \qquad (4a)$$

and, $$N_c = cot\phi(N_q - 1) \qquad (4b)$$

The intercepts of Eq. (1) will be equal to the bearing resistance, P_{bo}, when the normal stress is taken equal to zero. Eq. (3) then reduces to:

$$\mathrm{P_{bo} = nwd \cdot CN_c} \qquad (5)$$

With little or no overburden pressure, the bearing resistance will be a function of the soil's cohesion intercept, C. Accordingly, the initial part of these curves (at low normal stresses) are not necessarily linear. The intercepts deduced using Eq. (4b) are summarized in Table 2 and are shown in Figure 7. As shown in this figure, the calculated intercepts are different from the regression (best-fitting) intercepts which could be an indication of the nonlinearity at the initial parts of the curves.

TABLE 2. Calculation of passive resistance intercepts using the equation by Peterson and Anderson (1980)

Reinforcement	nwd, in^2	Punching		Bearing capacity	
		P_{bo}(lb/ft)	τ_o (psi)	P_{bo}(lb/ft)	τ_o (psi)
Stratagrid	9.39	208.7	0.24	377.6	0.44
UX-1500	5.01	111.4	0.13	201.4	0.23
Ux-1700	8.81	195.8	0.23	354.2	0.41

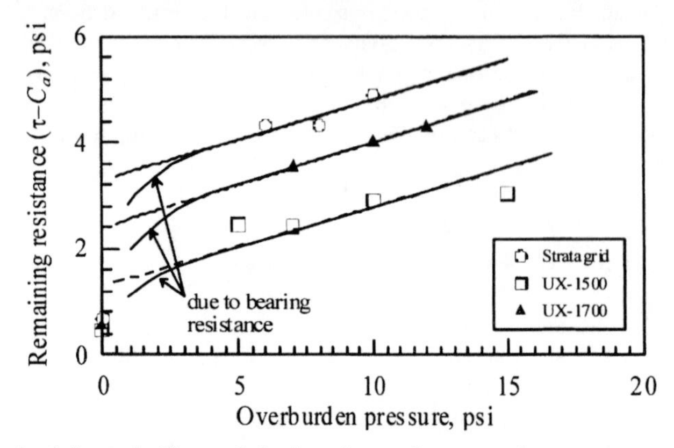

FIG. 7. Anticipated effects of the bearing resistances of transverse members on the remaining resistances for geogrids

CONCLUSIONS

This paper demonstrated the use of load-displacement curves measured at different points along the pullout strips to deduce the soil-reinforcement interface shear strength parameters. The measured load-displacement curves were used to produce the deformations in the reinforcements and the loads required to overcome the interface binding (adherence) for the reinforcement segment bounded between the two measurement points. The interface shear strength parameters are the apparent parameters since they were evaluated based on the total area of the strip. The results indicated that the apparent interface adhesion was generally independent of the level of stress of the test. The results of the tests also indicated that the contributions of the bearing resistances of the transverse members to the pullout resistances could be significant, especially at low confining stresses.

ACKNOWLEDGEMENTS

The work presented herein is based on a study funded by the Louisiana Department of Transportation- Louisiana Transportation Research Center (DOTD-LTRC). The DOTD and the LTRC are acknowledged for funding and facilities. The comments and suggestions of Mark Morvant, Pavement and Geotechnical Administrator are gratefully acknowledged. This work is based on an earlier research conducted by Dr. Khalid Farrag.

REFERNCES

Abu-Farsakh, M., and Almohd, I. A. (2004). "Evaluation of Interface Interaction between Geosynthetic Reinforcements and Cohesive Backfill". A Draft report submitted to the Louisiana Department of Transportation and Development, Louisiana Transportation Research Center, Baton Rouge, Louisiana.

Farrag, K., and Morvant, M. (2004). "Evaluation of Interaction Properties of Geosynthetics in Cohesive Soils: Lab and Field Pullout Tests," Report No. FHWA/LA.04/380, Louisiana Transportation Research Center, Baton Rouge, , Louisiana, 95 p.

Jewell, R. A., Milligan, G. W., Sarsby, R. W., and Dubois, D. (1984). "Interaction between Soil and Geogrids," *Proceeding from the Symposium on Polymer Grid Reinforcement in Civil Engineering*, London, England, pp. 18 – 30.

Mohiuddin, A. (2003). Analysis of Laboratory and Field Pull-out Tests of Geosynthetics in Clayey Soils, *MSc. Thesis*, Louisiana State University, Louisiana, .

Peterson, L.M. and Anderson, L.R. (1980). "Pull-out Resistance of Welded Wire Mats Embedded in Soil," *Master of Science Thesis*, Utah State University, Logan, Utah.

LAKE MICHIGAN BLUFF DRAINAGE EVALUATION FOX POINT AND OAK CREEK, WISCONSIN

Jeffrey Scott Miller, P.E.[1], Member, Geo-Institute, ASCE

ABSTRACT: Stability of soil slopes generally depends on gravitational forces of the slope geometry, soil characteristics, and subsurface water. Improving the stability of soil slopes requires reducing the gravitational forces, increasing the strength of the soil, or lowering the water level; individually or in combination. Gravitational forces can be reduced by lowering the slope height and reducing the slope angle. The strength of the soil can be increased by such methods as, among others, densification and modification in-place, internal reinforcement, and replacement. The water internally within a soil slope can be lowered by reducing water infiltration, or drainage, or both. Implementation of any of these improvement methods may be either difficult or not feasible due to space restrictions or regulations. A water drainage method installed internally within the natural soil slopes along the Lake Michigan shoreline in Milwaukee County, Wisconsin is being evaluated in an effort to improve the bluff slope stability with less disruption in comparison to other improvement methods. The drainage method being evaluated uses commercially available wicks, installed with directional drilling methods. To date, the study results show lowered water levels and a calculated improvement in the stability of the soil slope stability.

INTRODUCTION

Bluff slope instability has been a continuing problem along the west shoreline of Lake Michigan in Wisconsin. Public and private studies have credited the instability to two main sources: Lake Michigan wave erosion on the bluff toe, and water internally within the bluff soil profile.

This evaluation, which focuses on one of these main sources, analyzes draining the water internally within the bluff soil profile in order to increase the bluff stability. Geotechnical engineering exploration and analyses have been performed. Water level observation instrumentation and drainage systems have been installed. The evaluation is

[1]Sr. Project Manager, Giles Engineering Associates, Inc., N8W22350 Johnson Dr., Waukesha, WI. 53186 jmiller@gilesengr.com

currently continuing with observations of drainage system performance. Two specific sites along the Lake Michigan bluff in Milwaukee County, Wisconsin are being evaluated; a private property in Fox Point, Giles Engineering Associates, Inc. (2005) and the southern portion of Milwaukee County Bender Park in Oak Creek, Giles Engineering Associates, Inc. (2004). The Oak Creek project is partially funded by a grant to Milwaukee County, Wisconsin by the Wisconsin Coastal Management Program.

FOX POINT SITE DESCRIPTION

The Fox Point, Wisconsin site is a private residence. The site is located on the Lake Michigan Bluff at North Barnett Lane. The distance between the dwelling and the bluff crest is about 80 to 90 feet (24 to 27 m.), and the property width is about 110 feet (34 m.). The topography of the yard between the dwelling and the bluff crest is relatively level. The yard is approximately 125 feet (38m.) above Lake Michigan, and the bluff face is at an approximate 1.5 H: 1V slope, based on Southeastern Wisconsin Regional Planning Commission (SEWRPC) Topography Maps (1986).

No visual evidence of bluff instability is present at the top of the bluff. The bluff face is currently densely vegetated with mature trees and underbrush. Visual evidence is present of some past soil erosion from internal bluff water seepage, and down-slope soil movement. Sandy soils were found exposed on the bluff face in one area about one-third of the bluff height above the lake. The bluff toe has no rip-rap, and has a sand beach.

OAK CREEK SITE DESCRIPTION

The Oak Creek Wisconsin site is within Bender Park, a public access park in the Milwaukee County Park system. The general area evaluated is along the Lake Michigan shoreline in the southern portion of Bender Park between Fitzsimmons Road (extended) at the north and Oakwood Road (extended) at the south. This southern portion of Bender Park is about one half mile long. It is currently unimproved, except for a shoreline rip-rap stone revetment. Tall grass and brush generally cover the top of the bluff. Two isolated and environmentally sensitive areas of trees also exist at the bluff crest. The bluff face is steep, with immature brush and grass vegetation in some areas, and barren, eroded, and slumped soil in most areas of the face.

Two specific areas of the bluff are being evaluated. The locations are approximately 600 feet (183 m.) and 2100 feet (640 m.) in distance along the bluff crest and north-northwest of Oakwood Road extended. The two areas are hereafter referred to as the Southern Area and Northern Area. Topographic information discussed below is shown on topographs, Aero-Metric, Inc. (2001), prepared for the Milwaukee County Department of Public Works. This topographic information is the most recently available information for the bluff. The bluff height above Lake Michigan is about 116 feet (35 m.). Surface topography west of the bluff varies gradually by about 20 feet (6 m.). No evidence of recent bluff stability problems is present in the ground surface at the top of the bluff. No cracks and depressions indicative of bluff soil sliding were present in the soil surface. The bluff face is at an approximate average 1.2 H: 1V slope, at the Southern Area, and about an average 1.8 H: 1V slope at the Northern Area. Close observation of the bluff face was not attempted due to the steepness and barren soil condition. Areas of past water seepage, soil flow, and erosion are evident from observation at the top and bottom of the bluff. Scarps from shallow depth slope failures are generally present immediately below the bluff crest. Several large volume scarps from probable deep seated rotational slope failures are also present. Water-tolerant type vegetation is present at some locations on the area behind the rip-rap revetment at the bluff toe.

FOX POINT SITE SUBSURFACE CONDITIONS

Two geotechnical engineering test borings were performed at the Fox Point site for the purposes of bluff slope stabilization by Giles Engineering Associates, Inc. (2003) for the property owner. Each test boring was drilled to a depth of 121± feet (37± m.) below the ground surface near the bluff crest, east of the residential structure. The approximate test boring locations are indicated on the Test Boring Location Plan in Figure 1 below.

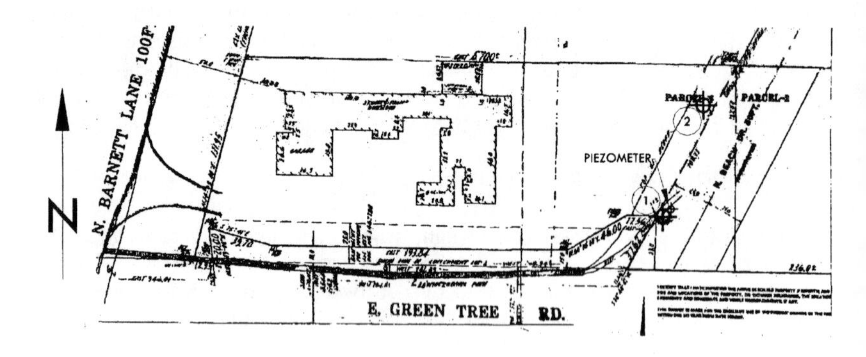

Fig. 1. Test Boring Location Plan Fox Point Site
(1.0 ft. = 0.3 m.)

A continuous auger core sampling was performed for each test boring. Laboratory tests performed consisted of natural moisture content, in place density, Atterberg Limits, and triaxial shear strength.

The subsoils encountered at the test boring locations generally consist of silty clay (Ozaukee Till overlying New Berlin Till) to at least the maximum depths explored. The silty clay is brown in color and has fissures to depths ranging from 9± to 13± feet (3± to 4± m.), and is gray below. A boulder was encountered at 7± feet (2± m.) in depth at Test Boring No. 2. Layers and seams of very fine sand, silt and clay were encountered between 63± and 100± feet (19± and 30± m.) in depth at Test Boring No. 1, and between 73± and 110± feet (22± and 34± m.) at Test Boring No. 2. Underlying soils encountered at the test borings consist of silty clay with sand and gravel, and sandy clay to the maximum depths explored.

Free water accumulated after the completion of drilling in Test Boring Nos. 1 and 2 at depths of 73± and 33± feet (22± and 10± m.), respectively, but the water levels are considered to have been affected by a bore hole cave-in. Wet sandy and silty soils were encountered at depths of 64± and 81± feet (20± to 25± m.) in depth at Test Boring Nos. 1 and 2, respectively. The groundwater table level at the time of the test boring exploration was estimated to be at a depth of 63± feet (19± m.) below the ground surface at the test boring locations.

OAK CREEK BENDER PARK SITE SUBSURFACE CONDITIONS

For this evaluation, four test borings were performed to characterize the subsoil profile and four pneumatic sensor piezometers were installed for recording the water level in the soils by Giles Engineering Associates, Inc. (2004). The test borings are numbered 1, 1A, 2, and 2A. Test Boring Nos. 1 and 1A are located in the Southern Area, and Nos. 2 and 2A are located in the Northern Area. The approximate locations are shown in Figure 2. Test Boring Nos. 1 and 2 were each drilled to a depth of 120± feet (37± m.) below the ground surface, and located near the bluff crest. Test Boring Nos. 1A and 2A were each drilled to a depth of 81± feet (25± m.) below the ground surface, and located approximately 50 feet (15 m.) west of Test Boring Nos. 1 and 2 respectively. Test Boring Nos. 1 and 2 were drilled with a continuous auger core sampling. Test Boring Nos. 1A and 2A were drilled with conventional Standard Penetration Test (SPT) (ASTM D-1586) sampling, which was performed in Test Boring No. 1 between 84± and 106± feet (26± and 32± m.) in depth where sandy soils were encountered. Laboratory tests performed consist of natural moisture content, in place density, Atterberg Limit, and triaxial shear strength.

The piezometers are numbered PZ-1, PZ-1A, PZ-2, and PZ-2A. The piezometers were installed in boreholes located about 5 to 10 feet (2 to 3 m.) away from the similarly numbered test borings, at the approximate locations shown in Figure 2. Soil sampling was not performed during borehole drilling for the piezometer installations, except for piezometer PZ-2.

The approximate test boring and piezometer locations are indicated on the Test Boring and Wick Locations plan (Figure 2).

Fig. 2. Test Boring and Wick Locations Oak Creek Site
(1.0 ft. = 0.3 m.)

The subsoils encountered at the test boring locations generally consist of silty clay (Oak Creek Till) to at least the maximum depths explored. The silty clay is brown in color and has fissures to depths ranging from 10± to 13± feet (3± and 4± m.), and is gray below. Coarse gravel or a cobble was encountered at 25± feet (8± m.) in depth at Test Boring No. 1A. Layers and seams of clay, silt, and very fine sand were encountered between 36± and 93± feet (11± and 28± m.) in depth at Test Boring No. 1, and below 42± feet (13± m.) at Test Boring No. 1A. At Test Boring No. 2, layers and seams or lenses of silty clay to clay and silt were encountered between 78± and 114± feet (24± and 35± m.) below the surface. Underlying soils encountered at the test borings consist of silty clay with sand and gravel, and sandy clay to at least the maximum depths explored.

Free water was encountered at depths of 60± and 45± feet (18± and 14± m.) below the surface at Test Boring Nos. 1 and 2, respectively, but did not accumulate after completion of drilling. Free water was not encountered during drilling and did not accumulate after drilling in the other test borings.

FOX POINT SITE DRAINAGE WICK INSTALLATION

Six drainage wicks were installed at the site, and completed by approximately March 2004. Directional drilling with thick viscosity slurry drilling fluid techniques were used to install a temporary flexible polymer pipe with a drainage wick attached. The drainage wick product is commercially available. The wick is a polymer plastic core with a bonded filter and separation geotextile. It is typically manufactured for consolidation-induced subsurface water drainage. The drilling insertion locations were at the top of the bluff, adjacent to the residential structure, and extended to the bluff face near the toe of the bluff. After the drilling reached the bluff face, the temporary polymer pipe was extracted, leaving the wick in place. The drilling fluid was also left in place, but reverted to a less viscous fluid and was discharged by gravity. The locations of the drainage wicks and their vertical and horizontal tracts are shown on the sketches by Underground Specialists, Inc. (2004) in Figures 3 and 4.

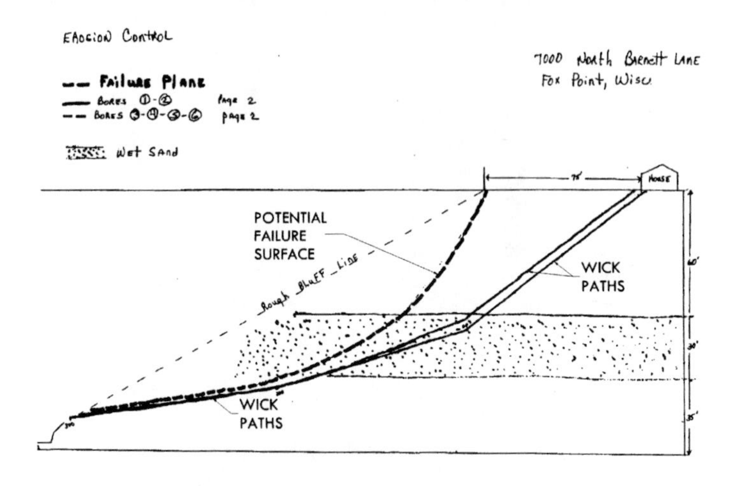

Fig. 3. Drainage Wick Installation Fox Point Site
(1.0 ft. = 0.3 m.)

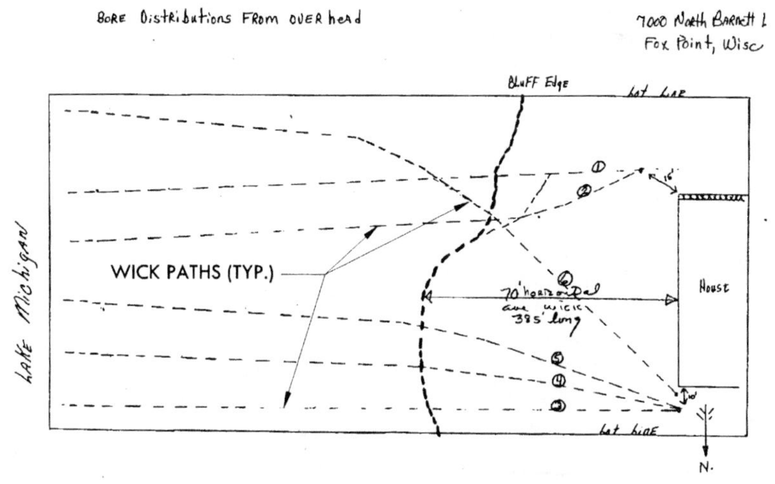

Fig. 4. Drainage Wick Installation Fox Point Site
(1.0 ft. = 0.3 m.)

Discharge of water from one of the wicks is shown in Figure 5. This photograph was taken about one week after this wick was completed.

Fig. 5. Drainage From Wick Fox Point Site

OAK CREEK SITE DRAINAGE WICK INSTALLATION

Initial installation of drainage wicks was done by approximately July 2004 at the Northern Area of the Oak Creek Bender site. Directional drilling techniques, and the drainage wicks product are the same as used at the Fox Point site. However, installation problems were encountered with these initial installations, which resulted in mostly ineffective wick performance to date. These wicks were installed at the bluff toe location, with a slight upward angle, but resulted in less than 40± feet westward penetration into the bluff.

The Southern Area wick drains were installed and completed by approximately August 2004. The installation was generally more successful than the wick installation in the Northern Area. The Southern Area wick drains were installed from the top of the bluff at a descending angle and eastward, and with an exit at the bluff toe. Subsequently, one additional wick drain was installed at the Northern Area and completed by approximately September 10, 2004 from the top of the bluff with an exit at the toe. The locations of the drainage wicks and their vertical and horizontal tracts are shown on the sketch by Underground Specialists, Inc. (2004) in Figure 6.

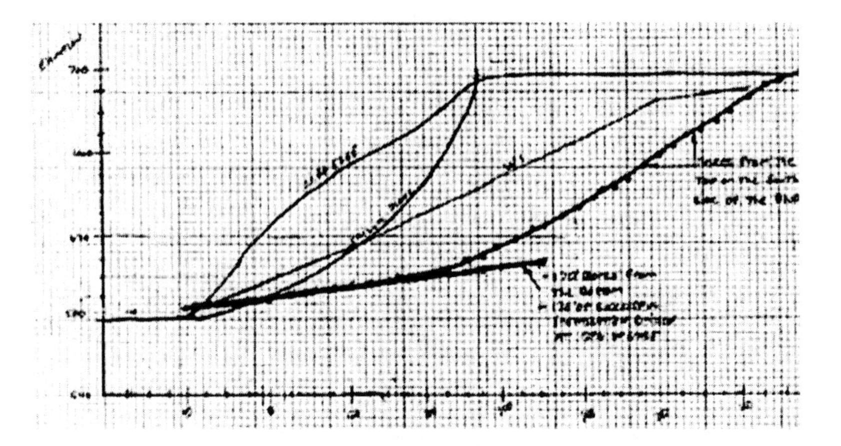

Fig. 6 Drainage Wick Installation Oak Creek Site
(1.0 ft. = 0.3 m.)

FOX POINT SITE GROUNDWATER LEVEL RECORDINGS

A piezometer was installed in a dedicated borehole at the site by Giles Engineering Associates, Inc. (2005), after the wick drains were in place. The piezometer is located near Test Boring No. 1, shown on the Boring Location Plan (Figure 1). The piezometer is a pneumatic sensor. It was placed inside the borehole with the sensor surrounded by a nominal 1 foot long filter sand pack, and with a bentonite clay seal and borehole backfill above the sand pack. The sensor was placed at a depth of 85 feet (26 m.) below the surface within a wet silt seam. The groundwater levels (piezometric level) within the bluff were measured by determining the piezometeric head above the elevation of the pneumatic piezometer sensor from the pressure of the water at the piezometer sensor. A pneumatic piezometer was selected for determining the groundwater levels because the response time between groundwater level changes and measurement of the levels in clay is almost instantaneous. This response time is considered to be an advantage in this analysis. The response time is about 42 days or longer for 2 inch (5 cm.) diameter standpipe-type observation wells. Those types of wells were not selected to measure the groundwater levels since a shorter response time is considered necessary to evaluate groundwater level changes.

Piezometric pressure readings and calculated approximate groundwater levels of the piezometer recorded since installation are shown in Table 1.

TABLE 1. Piezometer Record
(1.0 ft. = 0.3 m.; 1.0 psi = 6.9kN/m^2)

Ground Surface Elev. (feet)	Sensor Elev. (feet)	Date of Recording	Pressure (psi)	Water Level Elev. (feet)
705±	620±	6-13-2003	--	632 (a)
		3-25-2004	--	--(b)
		4-14-2004	--	--(c)
		5-18-2004	1.10	623
		5-24-2004	1.20	623 (d)
		6-30-2004	1.02	622
		8-18-2004	0.42	621
		12-6-2004	0.14	620
(a) Water level during drilling of nearby Test Boring No. 1 (b) Wick installation completed (c) Piezometer installed (d) Recording after heavy rainfall period				

The initial level reading shown is the level encountered in the nearby Test Boring No. 1 while it was being drilled in June 2003 about 9 months before drainage wick installation. Subsequent records are measurements of the water levels detected by the piezometer sensor.

The groundwater levels measured since piezometer installation are considered to represent a groundwater level condition lower than encountered in June 2003 when Test Boring No.1 was drilled. By December 6, 2004, the water level decreased by about 12 feet (4 m.) to about the piezometer sensor elevation.

Most of the water level decrease occurred between wick completion and the initial piezometer recording, and the rate of water level decrease has slowed. The slower rate of water level decrease is considered due to several factors. The water level hydraulic head on the wick drains has decreased by the way of water drainage through the wicks. Also, precipitation amounts since wick installation may have had an effect on the rate. Continued water level monitoring will enable a study on the effects of precipitation.

Recording was performed on May 24, 2004 to compare with the previous recording on May 18, 2004, since heavy rainfall occurred during the time between the dates of these two recordings. A water level pressure increase of about 0.1 pounds per square inch (0.7 kN/m^2), a rise of about 3 inches (8 cm.) in the water level) occurred between May 18, and May 24, 2004 based on the piezometer readings.

The Southeastern Wisconsin area received significant rainfall in May 2004. The recent *Newsletter*, Southeastern Wisconsin Regional Planning Commission (2004) describes the rainfall amounts during the 10 day period between May 14 to 23 as having a 1% probability of occurrence for most of the Milwaukee County area. The rainfall amounts during the first 24 days of May 2004 are described as having a probability of occurrence range between less than 0.2% to 0.5%. An official precipitation rain gage at the Bayside Middle School near the Fox Point site recorded a total rainfall of 10.86 inches (27.58 cm.) for this period.

OAK CREEK SITE GROUNDWATER LEVEL RECORDINGS

Four pneumatic piezometers were installed at the Oak Creek Site. Two piezometers numbered PZ-1, PZ-1A were installed at the Southern Area on October 1, 2003, and two piezometers numbered PZ-2 and PZ-2A in the Northern area on October 7, 2003. The piezometers were installed in boreholes located about 5 to 10 feet (2 to 3 m.) away from the similarly numbered test borings. The piezometers sensors and the borehole installation method are the same as used at the Fox Point site. The sensors were placed at depths of 80, 60, 75, and 50 feet (24, 18, 23, and 15 m.) below the surface at locations PZ-1, PZ-1A, PZ-2, and PZ-2A, respectively.

Two pre-existing observation wells are located about 100 feet (30 m.) north of Test Boring Nos. 2, 2A and piezometers PZ-2 and PZ-2A. The wells are nominal 2 inch (5 cm.) diameter standpipe-type wells. The bottom of the wells were determined to be 22± feet (7± m.) and 100± feet (30± m.) below the ground surface. Their bottom depths and periodic measurements of the water levels within the wells were measured for this evaluation with an electric water-level sensor.

Piezometric head readings of the piezometers recorded since installation are shown in Tables 2 and 3. Water level recordings of the two standpipe observation wells, labeled STS 1 and STS 2 are also included for reference purposes.

TABLE 2. Piezometer Record Oak Creek WI Bender Park Site
(1.0 ft. = 0.3 m.; 1.0 psi = 6.9kN/m^2)

Piezometer Location	PZ 1	PZ 1A	PZ 2	PZ 2A	STS 1	STS 2
Ground Surface Elev. (feet)	697±	698±	696±	698±	695±	695±
Sensor Elev. (feet)	617±	638±	621±	648±	--	--
	Pressure (psi) or Depth (feet)					
10-2-2003	8.01	1.00	--	--	49.0	8.5
11-21-2003	7.94	3.84	14.96	12.37	--	--
12-24-2003	8.05	4.15	14.79	13.37	48.2	12.9
5-18-2003	10.12	6.10	16.80	15.60	46.9	0.5
5-24-2004	10.30	6.40	17.10	15.90	46.7	1.0
7-8-2004	10.75	7.05	17.30	15.56	45.4	2.1
8-4-2004	5.45	4.45	17.60	15.25	--	--
8-18-2004	4.14	2.14	17.45	15.14	--	--
9-23-2004	2.55	0.92	16.94	14.16	46.0	6.0
10-21-2004	1.88	0.76	16.11	13.60	--	--
12-17-2004	1.15	0.69	15.30	12.95	--	--

TABLE 3. Piezometer Record Oak Creek WI Bender Park Site
(1.0 ft. = 0.3 m.; 1.0 psi = 6.9kN/m^2)

Piezometer Location	PZ 1	PZ 1A	PZ 2	PZ 2A	STS 1	STS 2
Ground Surface Elev. (feet)	697±	698±	696±	698±	695±	695±
Sensor Elev. (feet)	617±	638±	621±	648±	--	--
	Water Level Elevation (feet)					
10-2-2003	635	640	--	--	646	687
11-21-2003	635	647	656	677	--	--
12-24-2003	636	648	655	679	647	682
5-18-2003	640	652	660	684	648	695
5-24-2004	641	653	660	685	648	694
7-8-2004	642	654	661	684	650	693
8-4-2004	630	648	662	683	--	--
8-18-2004	627	643	661	683	--	--
9-23-2004	623	640	660	681	649	689
10-21-2004	621	640	658	679	--	--
12-17-2004	620	640	656	678	--	--

The initial level readings for piezometers PZ-1 and PZ-1A were obtained a short time after installation and may not be accurate due to the residual water pressures caused by the installation. Subsequent PZ-1 and PZ-1A records and all of the PZ-2 and PZ-2A records are considered valid representations of the groundwater levels at the time of recording.

The groundwater levels recorded indicate a lower groundwater elevation at the piezometers closest to the bluff crest and Lake Michigan. This is considered reasonable since groundwater flow from the higher elevation within the bluff to the lower lake level elevation is logical.

The groundwater levels measured between piezometer installation and wick drainage installation represent a relatively low groundwater level condition. An increase in regional precipitation is generally predicted by the historically cyclical levels of the Great Lakes relative to the recent past near record low Lake Michigan water level. The Lake Michigan water level is expected to rise in the next several years, based on publicly recorded lake level statistics.

Recording was performed on May 24, 2004 to compare with the previous recording on May 18, 2004, since heavy rainfall occurred during the time between the dates of these two recordings. A water level pressure increase at all of the piezometers of about 0.3 pounds per square inch (2.1 kN/m^2), a rise of about 8 inches (20 cm.) in the water level) occurred between May 18, and May 24, 2004 based on the pneumatic piezometer readings. Water levels decreased in the standpipe piezometers during the same period.

The wick drains in the Southern Area were installed during July 2004 and completed prior to the readings taken on August 4, 2004. Water level decreases of 12± and 6± feet (4± and 2± m.) were recorded on August 4 at piezometers PZ1 and PZ1A from the pre-wick installation recordings. The initial and unsuccessful wick drains in the Northern Area were installed during July 2003. Water level recordings at piezometers PZ2 and PZ2A indicated water level increases after the May 2004 heavy rainfall events until the successful installation of the one wick drain from the top of the bluff completed on September 10, 2004. The water levels at the piezometers decreased by about 2± and 4± feet (0.6± and 1.2± m.) at piezometers PZ2 and PZ2A respectively on September 23, 2004, verses the previous high level readings. At the time of the most recent recording, the water levels have decreased at piezometers PZ1 and PZ1A by 22± and 14± feet (7± and 4± m.) respectively since wick drain installation was completed at the Southern Area. At the time of the most recent recording, the water levels have decreased at piezometers PZ2 and PZ2A by 6± and 5± feet (2± and 2± m.) respectively since wick drain installation was completed at the Northern Area. Most of the water level decrease occurred between wick completion and the initial piezometer recording, and the rate of water level decrease slowed. However, water seepage from the wick drain discharges is occurring, as shown in Figure 7 taken at the time of the most recent recording in December, 2004.

Fig. 7. Drainage From Wick Oak Creek Site

FOX POINT SITE SLOPE STABILITY ANALYSES

Slope stability analysis calculations were performed for the pre-drainage wick installation condition of the Fox Point site slope. A factor of safety value of 1.0 was determined by the stability analysis, which indicates a possible occurrence of deep rotational slides. No evidence of recent bluff stability problems was present in the ground surface at the top of the bluff. No cracks and no depressions indicative of bluff soil sliding were present in the soil surface, at the time of the analysis. Mature and relatively vertical trees are present on the bluff face. All of these are indicators of deep-seated bluff face stability.

The stability analysis used the subsurface conditions and soil properties determined from the test borings and laboratory testing. The properties are shown in Table 4 below. The water level used in the analysis was at an assumed depth of 63± feet (19± m.) below the ground surface at the test boring locations, and at the bottom of the bluff at the toe elevation. The toe of the bluff was used as the groundwater level at the toe location since no water seepage from the bluff face was visually apparent. The bluff height and horizontal distance from the bluff toe, and the bluff face topography were obtained from the *Topography Maps,* Southeastern Wisconsin Regional Planning Commission (1986).

Another stability analysis was performed for the slope with the December 6, 2004 recorded water level at the piezometer location. With a water level depressed by 12± feet (4± m.) in elevation, the factor of safety increases to about 1.34, an increase of about 34%.

TABLE 4. Soil Properties Fox Point Site
(1.00 pcf = 0.16 kN/m^3; 1.00 psf = 0.05 kN/m^2)

Soil	Unit Weight (pcf)	Effective Cohesion (psf)	Effective Friction Angle (degrees)
Silty Clay	135	850	27
Sands and Silts	135	0	39
Silty Clay	130	360	29
Sandy Clay	145	0	35

OAK CREEK SITE SLOPE STABILITY ANALYSES

Slope stability analysis calculations were performed for the pre-drainage wick installation condition of the Oak Creek Bender Park Northern and Southern Area slopes. A factor of safety value of 1.0 and 1.17 was determined by the stability analyses for the bluff slope at the respective Southern and Northern Areas studied, which indicates a possible occurrence of deep rotational slides.

Bluff topographic profiles determined in April 2001, Aero-Metric, Inc. (2001) were used for the analyses, since they are the most recently available topographic information. Some change in the bluff face shape and slope angle may have occurred since 2001, as evidenced by the barren, eroded, and slumped soil in most areas of the face. The factors of safety results of the stability analyses are considered representative of the bluff face topography in 2001, and are considered relative to the present conditions.

Slope stability analysis calculations were based on the subsurface conditions and the engineering properties of the subsoils determined by field and laboratory tests. The properties are shown in Table 5 below. The groundwater level elevations relative to the horizontal distances within the bluff were estimated by interpolation between the levels measured at the piezometer locations and near the toe of the bluff.

Two other stability calculations were made to evaluate the effect of the lower groundwater level recently recorded at the piezometer locations. With the water level depressed, the factor of safety values are calculated to be 1.21 and 1.11 at the Northern Area and Southern Area respectively. The Factor of safety values increased by about 3 to 11 percent at the Northern and Southern Areas, respectively.

TABLE 5. Soil Properties Oak Creek Site
(1.00 pcf = 0.16 kN/m^3; 1.00 psf = 0.05 kN/m^2)

Soil	Unit Weight (pcf)	Effective Cohesion (psf)	Effective Friction Angle (degrees)
Silty Clay	138	310	30
Clay	129.8	570	29.5
Sand	120	0	32
Silty Clay	132.6	380	28

CONCLUSIONS

The evaluations of drainage of the Lake Michigan bluff slopes at two Milwaukee County Wisconsin sites have the following conclusions:

- The installations of drains have been effective in lowering the water levels internally within the bluffs.
- The lower water levels have increased the slope stability factor of safety. Continued drainage will increase the factors of safety.
- Pre-fabricated polymer wicks installed with horizontal directional drilling methods are successful drains and successful for stabilizing slopes.
- Wick drain installation with horizontal directional drilling methods is a feasible tool for slope stabilization where other stabilizing methods are not feasible due to space restrictions or regulations.

ACKNOWLEDGEMENTS

The author acknowledges Giles Engineering Associates, Inc., Edward D. Gillen Co., and Underground Specialists, Inc. for giving permission to publish this data. The author also acknowledges the employees of Giles Engineering Associates, Inc. for the exploration and laboratory work, and Underground Specialists, Inc. for the wick installations. Special thanks go to Mr. Gary Jackson of Edward D. Gillen Co.

REFERENCES

Aero-Metric, Inc. (2001). *Topographic Mapping of Bender Park* Milwaukee County, Wisconsin.

Giles Engineering Associates, Inc. (2003). *Geotechnical Engineering Exploration and Lake Michigan Bluff Stability Analysis* Residence, 7000 N. Barnett Lane, Fox Point, Wisconsin.

Giles Engineering Associates, Inc. (2004). *Geotechnical Engineering Exploration and Lake Michigan Bluff Stability Analysis* Bender Park, Oakwood Road, Oak Creek, Wisconsin.

Giles Engineering Associates, Inc. (2005). *Water Level Observations*, Residence, 7000 N. Barnett Lane, Fox Point, Wisconsin.

Southeastern Wisconsin Regional Planning Commission. (1986). *Topography Maps* NE ¼ Section 21 T8N R22E, and NW ¼ Section 21 T8N R22E

Southeastern Wisconsin Regional Planning Commission. (2004). *Newsletter* Vol. 42, No.1.

Underground Specialists, Inc. (2004). *Completion Report for the Bender Park Wick Drain Project,* Oak Creek, Wisconsin.

Underground Specialists, Inc. (2004). *Sketches*, Fox Point, Wisconsin.

APPLYING SEPARATE SAFETY FACTORS TO END-OF-DRIVE AND SET-UP COMPONENTS OF DRIVEN PILE CAPACITY

Van E. Komurka[1], P.E., Member, ASCE, Charles J. Winter[2], P.E., Member, ASCE, and Steven G. Maxwell[3], P.E.

ABSTRACT: A driven pile's long-term capacity is often the sum of two components: end-of-initial-drive capacity, and set-up (time-dependent capacity increase). Incorporating set-up into pile design and installation procedures has many economic advantages (potentially millions of dollars), and is increasing in acceptance and application among designers. For a number of reasons, it may be desirable to apply separate (different) safety factors to the end-of-initial-drive capacity and set-up components. This approach is particularly well-suited to load and resistance factor design, which is anticipated to be used on all new bridge designs after 2007. The analytical approach to applying separate factors of safety is presented, and its application is illustrated in a case history from the Marquette Interchange project in Milwaukee, Wisconsin. In the case history, separate safety factors for end-of-initial-drive capacity and set-up were selected based on the results of a design-phase pile test program, and set-up safety factors varied with pile diameter.

INTRODUCTION

The use of load-factor design procedures is not new to the structural community, their use for bridge superstructure design is expanding rapidly. However, portions

[1]Vice President, Wagner Komurka Geotechnical Group, Inc., W67 N222 Evergreen Boulevard, Suite 100, Cedarburg, Wisconsin, 53012, komurka@wkg2.com.
[2]Geotechnical Engineer, Wagner Komurka Geotechnical Group, Inc., W67 N222 Evergreen Boulevard, Suite 100, Cedarburg, Wisconsin, 53012.
[3]Geotechnical Engineer, Wisconsin Department of Transportation, Transportation District 2, 141 N.W. Barstow Street, P.O. Box 798, Waukesha, Wisconsin, 53187-0798.

of substructure design (e.g., geotechnical pile capacity) is traditionally based on allowable stress methods. Applying separate safety factors to account for differing uncertainties among pile design components has received increased interest in recent years. This method shares many principles with the Load and Resistance Factor Design ("LRFD") approach.

A pile under load can fail for lack of structural capacity, or for lack of geotechnical capacity (by unacceptable penetration into the ground). Structural failures of piles meeting the specified installation criteria are rare, and are not further discussed. A driven pile's long-term geotechnical capacity is often the sum of two separate and distinct components: its end-of-initial-drive ("EOID") capacity, and set-up. This paper deals with these two components of geotechnical capacity.

The methods used to estimate EOID capacity and set-up may differ. Capacity analysis methods may differ by component (EOID capacity vs. set-up), and by when they are performed (during design vs. during construction). The results of such analyses may have different degrees of uncertainties, or may yield different ranges of results. As part of the overall design process, it is important that the foundation designer qualitatively assess the reliability of the geotechnical design parameters. For these reasons, it may be desirable to apply separate safety factors to EOID capacity and set-up.

This paper describes some of the methods available to evaluate EOID capacity and set-up (during both design and construction), and potential reasons for applying separate safety factors to these two capacity components. An analytical approach is presented for application of different safety factors to EOID capacity and set-up. A case history is presented which describes the analytical approach, discusses the role design-phase test results played in the selection of separate safety factors, and illustrates the application in developing installation criteria.

SOIL/PILE SET-UP

It is known that after installation, pile capacity may increase with time. This capacity increase is known as set-up, and was first mentioned in the literature in 1900 by Wendel [Long et al., 1999]. Set-up has been documented in fine-grained soils in most parts of the world [Soderberg, 1961], and has been demonstrated to account for capacity increases of up to 12 times the initial value [Titi and Wathugala, 1999]. Set-up rate and magnitude is a function of a combination of a number of factors [Komurka et al., 2003a; Samson and Authier, 1986], the interrelationship of which is not well understood. Set-up is primarily attributable to an increase in shaft resistance [Axelsson, 2002; Bullock, 1999; Chow et al., 1998].

Incorporating set-up into pile design and installation has many advantages, and is increasing in acceptance and application among designers and agencies. By incorporating set-up into design, it may be possible to increase allowable pile loads, and to reduce: the number of piles, pile lengths (and potentially splices), pile sections (use smaller-diameter or thinner-walled pipe piles, or smaller-section H-piles), driving equipment size (use smaller hammers and/or cranes), or installation time, all of which should result in lower costs. A number of projects have documented savings in the millions of dollars.

POTENTIAL REASONS TO APPLY SEPARATE SAFETY FACTORS

Pile capacity must exceed applied loads by a sufficient margin so that the foundation does not fail structurally or geotechnically. Safety factors are applied to pile capacities (resulting in allowable loads) to account for uncertainties in applied load (loads or loading conditions, load determination methods, foundation stiffness, thermal effects, etc.) and uncertainties in resistance to those loads (extent and quality of the site investigation program including field and laboratory testing, variability of subsurface conditions across the site, reliability of soil strength data, pile capacity evaluation methods, quality control procedures (including the ability to install the pile without structural defects and capacity verification measures), pile material properties, installation equipment performance, environmental effects, etc.).

Discussion of specific safety factor selection based on these factors is beyond the scope of this paper. Statistical methods can assess risk, and form the basis for the safety factors proposed by modern codes. Likins (2004) presents a review of several codes' recommended safety factors, showing most codes relate safety factor selection to the type and amount of capacity verification performed.

To estimate set-up, pile capacity requires evaluation, both at EOID and at some later time. There are a number of approaches to evaluating capacity at EOID and at some later time, each with its own associated limitations and uncertainties. For this, and other, reasons, it may be desirable to apply separate safety factors to these two capacity components.

Capacity Components' Evaluation Methods

The capacity determination methods used to evaluate EOID and longer-term capacity may have different associated uncertainties. For example, design-phase EOID capacity may use dynamic testing and CAse Pile Wave Analysis Program (CAPWAP®) analyses. Longer-term capacities (set-up) may be estimated using empirical formula, or extrapolated from relatively short-term static loading or restrike tests. In this case, EOID capacities may be considered to have less uncertainty than set-up. An awareness of relative uncertainties between EOID capacity and set-up evaluations should play a role in the decision to use, and the selection of, separate safety factors.

Design-Phase — End-Of-Initial-Drive Capacity Determination

Static Analysis — Static analysis methods can be categorized as analytical methods which use soil strength/relative density properties to determine pile capacity, and so do not rely on any pile driving data. A large number of static analysis methods are documented in the literature, with specific recommendations on the safety factor to be used with each method (although these recommended safety factors have routinely discarded the influence of the construction control method used to complement the static analysis computation). Most static analysis methods recommend a safety factor of 3. Piles whose designs are based solely on static analyses (albeit rare) might be installed to a minimum depth criterion. In comparison with the other methods described (with the possible exception of certain dynamic formulas), static analyses are typically considered to have the greatest degree of uncertainty.

Dynamic Formulas — Dynamic formulas are based on energy concepts relating energy applied by the hammer to work done by the pile penetrating the soil, and so rely on penetration resistance (pile set per blow) during driving to analyze capacity. The inadequacies of dynamic formulas have been known for a long time [Peck, 1942]. Dynamic formulas are fundamentally incorrect: the derivation of most formulas is not based on a realistic treatment of the driving system, the soil resistance is very crudely treated by assuming it is a constant force, and usually the pile is assumed to be rigid and its length is not considered. Regarding the actual safety factor obtained by using the Engineering News formula (a popular dynamic formula), Chellis (1961) noted that it ranged from 1/2 to 16, Sowers (1979) reported that it ranged from 2/3 to 20, and Rausche et al. (1996a) determined that it ranged from 0.6 to 13.1. While most formulas are typically considered to have less uncertainty than static analysis methods, dynamic formulas are considered to have relatively high uncertainties when compared to wave equation analysis and dynamic pile testing and analysis.

Wave Equation Analysis — Wave equation analysis offers a complete approach to the mathematical representation of a system consisting of hammer, cushion(s), helmet, pile, and soil, using an associated computer program for the convenient calculation of the motions and forces in this system after ram impact. The approach was developed by E.A.L. Smith (1960). After the rationality of the approach had been recognized, several researchers developed a number of computer programs. Although wave equation analysis can be used to evaluate a number of installation parameters (e.g., driving stresses), a primary application is to develop a bearing graph relating pile capacity to penetration resistance. Relatively speaking, this method has less uncertainty than either static analysis or most dynamic formulas. However, this method lacks direct measurement on a driven pile at the project site, and therefore is considered to have more uncertainty than dynamic pile testing and analysis.

Dynamic Pile Testing and Analysis — Dynamic pile testing methods use measurements of strain and acceleration taken near the pile head as a pile is driven. Among other things, these dynamic measurements can be used to estimate static pile capacity in the field during driving using the Case Method [Goble and Rausche, 1970; Goble et al., 1975; Rausche et. al., 1985]. Subsequent additional analysis of dynamic monitoring data may include performing a CAPWAP analysis (a rigorous numerical modeling technique) to refine capacity estimates, and to provide assessment of capacity allocation (toe resistance versus shaft resistance, and shaft resistance distribution) [Hussein et al., 2002; Likins et al., 1996; Likins and Rausch, 2004; Rausche et al., 1972, 1994, 1996b, 2000]. Since this method involves direct measurements on a driven pile at the project site, it is generally considered to have the least degree of uncertainty of the methods described.

Design-Phase — Set-Up Determination

To determine set-up, pile capacity must be evaluated at EOID and at some later time. EOID capacity is subtracted from longer-term capacity to determine set-up. Since set-up is the difference between EOID capacity and longer-term capacity, the accuracy of set-up so determined is sensitive to the accuracy of both EOID, and longer-term, capacity evaluations [Komurka, 2004]. Accordingly, set-up has greater uncertainty than either EOID or longer-term capacity.

Static Analysis — Some static analysis methods may have provision for incorporating set-up. Such provision may be in the form of inputting a set-up factor, a cohesive soil sensitivity, or a percentage strength loss during driving. These inputs, and the reliability of the method, may be based on soil type, field or laboratory testing results, published relationships, local experience, etc. If a static analysis method is empirically correlated to static loading test results, set-up may already be incorporated into the correlation since a static loading test cannot be performed instantaneously after driving (i.e., set-up occurs before the static loading test can be performed). Designers should fully understand the basis for, and the limitations and applicability of, a chosen static analysis method, particularly with respect to incorporating set-up into design. Even with such an understanding, such set-up evaluations are typically considered to have a high degree of uncertainty.

Empirical Relationships — A number of researchers have offered empirical relationships for predicting pile capacity with time if capacity at some initial time is known [Guang-Yu, 1988; Huang, 1988; Skov and Denver (1988); Svinkin, 1996; and Svinkin and Skov (2000)]. Such relationships are subject to a number of limitations (Komurka et al., 2003b), and should be used judiciously by designers with local experience correlating predictions to results. The relative uncertainty of empirical relationships' set-up predictions depends on how closely the project conditions emulate the conditions and assumptions on which the relationships were based.

Restrike Testing — Restrike testing involves redriving a pile with a pile driving hammer some time after installation to evaluate longer-term capacity. For dynamic formula and wave equation analysis, restrike testing penetration resistance is used to evaluate capacity. For Case Method estimates and CAPWAP analyses, dynamic measurements obtained during restrike testing are used to evaluate capacity. Because of set-up, mobilizing full capacity during restrike testing often requires a larger hammer (i.e., more impact force) than used for installation. Since restrike testing involves direct measurements on a driven pile at the project site, it can have a relatively low uncertainty, but its associated uncertainty depends on if full capacity is mobilized, type of restrike data obtained, and the method of analysis applied to the data.

Static Load Testing — Static loading tests have traditionally been the standard for evaluating pile capacity. If set-up is present, the capacity measured by a static loading test almost always includes a set-up component, since a static loading test cannot be performed instantaneously after driving. To evaluate set-up from a static loading test, EOID capacity must be subtracted from the static-loading-test-determined capacity (determining set-up distribution from a static loading test requires instrumentation to evaluate load transfer behavior). Accordingly, the relative uncertainty associated with set-up determination from a static loading test lies predominately with the EOID capacity evaluation.

Construction Phase — EOID Capacity and Set-Up Determination

Many of the same capacity determination methods used in the design phase can also be employed in the construction phase. However, uncertainties associated with each of these methods may differ from their use in the design phase, due to construction control procedures. For example, construction control procedures may include periodic EOID capacity evaluation using dynamic testing and CAPWAP analyses, with provision to modify installation criteria based on results. Longer-term capacities (set-up) may not be evaluated further during construction, or may be evaluated with relatively short-term restrike testing and extrapolated to longer-term capacity. In this case, EOID capacities may be considered to have less uncertainty than set-up.

Site Coverage

EOID capacity and set-up evaluation uncertainties may depend on the extent to which testing can characterize a site. For example, in a relatively small building footprint, a design-phase test program may characterize driving behavior and set-up to a greater extent than possible on a project of large plan area (e.g., an interchange). If relatively less testing coverage means more interpolation or extrapolation of results is required for design, and EOID capacity variations are accounted for by penetration resistance criteria, EOID capacities may be considered to have less uncertainty than set-up.

Variability of Results

Test results may indicate variable pile behavior (both EOID capacity and set-up), even between relatively proximate locations, or between apparently similar soil conditions. With penetration-resistance-based installation criteria, variations in EOID capacity are evidenced and accounted for during installation by variations in penetration resistance. Potential variations in set-up are much less discernable during driving. In this case, EOID capacities may be considered to have less uncertainty than set-up.

Application of Results

Test results may be applied to slightly different piles than from which the results were obtained (e.g., applying unit shaft resistance or unit set-up values obtained from 12.75-inch-diameter test piles to 14-inch-diameter production piles). Increased uncertainties in both EOID capacity and set-up are likely with such extrapolation.

Relative Contribution of Set-Up to Pile Capacity

The effort of deciding to use, selecting, and applying separate safety factors for EOID capacity and set-up may not be worthwhile if anticipated set-up provides a relatively small contribution to pile capacity. In this case, actual set-up significantly less than anticipated in design may result in a relatively minor reduction in the overall safety factor. Conversely, if anticipated set-up provides a relatively large contribution to pile capacity, actual set-up significantly less than anticipated in design may result in an unacceptable reduction in the overall safety factor. In this case, consideration should be given to applying a higher safety factor to set-up.

Compatibility With Load and Resistance Factor Design

It may be possible to incorporate applying separate safety factors to EOID capacity and set-up of driven piles into the Load and Resistance Factor Design ("LRFD") method, or other load-factor methods. Mechanisms for applying resistance factors to geotechnical pile capacities in LRFD have been documented by Liang and Nawari (2000), and Paikowsky (2004). Additional discussion of LRFD and/or other load-factor methods in the design and installation of deep foundations have been presented by Goble, et al. (1980), Likins (2004), and Long (2002). However, application of the LRFD concept to foundation design is not universally accepted, and doubts have been expressed about its use [Svinkin, 2003].

Acceptance of Set-Up in Design

A pile foundation designer's recommendation to incorporate set-up into driven pile design and installation may be received with reluctance or skepticism from others (an owner, a reviewing agency, design team members, etc.). For these others involved, it may be a first-time use or a relatively new approach, or set-up magnitudes or relative contributions may be high enough to foster reservations. In such cases, an appropriately high set-up safety factor may increase comfort levels so that incorporating set-up into design is acceptable. Incorporating set-up with a high safety factor is better than ignoring set-up completely.

INCORPORATING SEPARATE SAFETY FACTORS INTO PILE DESIGN

Overall Safety Factor

The uncertainties discussed in the previous section can be addressed using a load-factor procedure. This procedure shares the same philosophy as LRFD, in that the EOID and set-up components of long-term capacity are assigned separate safety factors respective of their relative uncertainty.

The relative contributions from EOID capacity and set-up to ultimate long-term capacity influence the affect of each component's uncertainties on the overall safety factor[4]. The overall safety factor can be determined starting with:

$$EOID + SetUp = ULTC \qquad (1)$$

where: EOID = EOID Capacity Component of Ultimate Long-Term Capacity
SetUp = Set-Up Component of Ultimate Long-Term Capacity
ULTC = Ultimate Long-Term Capacity

[4] "Ultimate capacity" is a misnomer, as capacity of the deep-foundation element (e.g., "bearing capacity," "uplift capacity," "shaft capacity," and "toe capacity") is the ultimate resistance of the element. It cannot be misunderstood, however, and aids avoiding confusion with allowable load, and so is used herein.

Dividing by respective safety factors yields:

$$\frac{EOID}{SF_{EOID}}+\frac{SetUp}{SF_{SETUP}}=\frac{ULTC}{SF_{OVERALL}}=Allowable\ \ Load \tag{2}$$

where: SF_{EOID} = Safety Factor Applied to EOID Component of Ultimate Long-Term Capacity
SF_{SETUP} = Safety Factor Applied to Set-Up Component of Ultimate Long-Term Capacity
$SF_{OVERALL}$ = Overall Safety Factor
Allowable Load = Allowable Pile Load

Cross-multiplying to determine a common denominator, and rearranging, yields:

$$SF_{OVERALL}=\frac{ULTC\times SF_{EOID}\times SF_{SETUP}}{EOID\times SF_{SETUP}+SetUp\times SF_{EOID}} \tag{3}$$

Inspection of Eq. 3 indicates that as a capacity component's relative contribution increases, the overall safety factor approaches that component's safety factor.

As discussed previously, uncertainty associated with a capacity component which has a relatively small contribution to long-term capacity has diminished effect on the overall safety factor. Consider two cases: the first in which set-up is anticipated to contribute relatively little (on the order of ten percent) to long-term capacity, and the second in which set-up is anticipated to contribute significantly (on the order of 70 percent) to long-term capacity. If only half the anticipated set-up actually occurred, the resulting actual overall safety factor would be affected much less for the first case than for the second case. Accordingly, consideration could be given to applying a lower set-up safety factor for the first case than for the second case.

Required End-Of-Initial-Drive Capacity

For installation criteria development, after a desired allowable load is selected, the EOID capacity to which the piles should be installed is of principal interest. Multiplying both sides of Eq. 2 by SF_{EOID} results in:

$$Req'd\ \ EOID+SetUp\times\frac{SF_{EOID}}{SF_{SETUP}}=Desired\ \ Allowable\ \ Load\times SF_{EOID} \tag{4}$$

The term {SetUp x (SF_{EOID} / SF_{SETUP})} in Eq. 4 is herein referred to as the "adjusted set-up." It should be noted that both sides of Eq. 2 could just as easily have been multiplied by SF_{SETUP}, in which case required EOID capacity would be adjusted, and allowable load would have been multiplied by SF_{SETUP}. It should also be noted that a pile's overall safety factor is not the factor by which the allowable pile load is multiplied in Equation 4, but instead is determined by Eq. 3. The sum of EOID capacity plus unadjusted (actual) set-up still equals the ultimate long-term capacity. Rearranging Eq. 4 results in:

$$Req'd\ \ EOID = (Desired\ \ Allowable\ \ Load) \times SF_{EOID} - SetUp \times \frac{SF_{EOID}}{SF_{SETUP}} \qquad (5)$$

With Eq. 5, required EOID capacity can be determined using the set-up magnitude used for design, the desired allowable pile load, and separate safety factors for EOID capacity and set-up. This methodology is demonstrated in the following case history example.

The quantity {(Desired Allowable Load) x SF_{EOID}} in Eq. 5 is merely a value from which adjusted set-up is subtracted to determine required EOID capacity. It should not be confused with the ultimate long-term capacity in Eq. 1. This distinction is illustrated in the case history.

CASE HISTORY EXAMPLE

Project Description

The Marquette Interchange project at the junction of interstate highways I-43 and I-94 near downtown Milwaukee, Wisconsin, is an $810-million interchange replacement project, is the largest in state history, and is currently the most-complex design underway in the country. Encompassing 80 acres, the project includes two million square feet of bridge decks, 19 bridge structures, supported by 265 substructures, and is classified by the Federal Highway Administration as a megaproject.

In the summer of 2003, a $2-million design-phase pile test program was performed, with emphasis on characterizing EOID capacity, set-up, and long-term capacity as functions of depth for use in design, installation, and production control of driven pile foundations. Eighty-nine test piles, consisting of 12.75-, 14-, and 16-inch-diameter closed-end steel pipe, were driven. A detailed description of the test program is beyond the scope of this paper. To characterize set-up, dynamic monitoring using a Pile Driving Analyzer® ("PDA") [Goble et al., 1975; Hannigan et al., 1997; Pile Dynamics, Inc., 1998; Rausche et al., 1985] was performed during installation, and during subsequent restrike testing. CAPWAP analyses were performed on dynamic monitoring data from representative EOID and beginning-of-restrike ("BOR") blows.

Separate Safety Factors

Design-phase pile test program results were reviewed to select appropriate separate EOID capacity and set-up safety factors. At review time, it was desired to provide the design option of using 12.75-, 14-, or 16-inch-diameter steel pipe piles of various wall thicknesses, with allowable loads of up to 150, 200, and 250 tons, respectively.

For EOID capacity predictions, two Case Method equations, RA2 and RX9, were used to evaluate capacity vs. penetration depth during installation. The RA2 and RX9 capacities were compared to one another, and also to CAPWAP results.

For set-up prediction, unit set-up distributions (discussed subsequently) among piles were compared. This comparison indicated that unit set-up distributions varied, sometimes significantly, among relatively proximate piles. The effect of this variation (and potential overprediction of set-up) on reducing the actual overall safety factor was evaluated for a number of potential design cases. Detailed discussion of this evaluation is beyond the scope of this paper. It was determined that the extent to which the actual overall safety factor is reduced by potential overprediction of set-up is a function of the desired overall safety factor, allowable pile load, pile diameter, and set-up's relative contribution to long-term capacity.

The separate safety factors adopted by the design team and Owner as a result of this evaluation are presented in Table 1.

TABLE 1. Safety Factors Used for Design and Installation.

Pile Diameter, inches	Capacity Component	
	EOID	Set-Up
12.75	2.25	2.50
14	2.25	2.50
16	2.25	2.75

Set-Up Used for Design

For this project, CAPWAP results were used to estimate unit shaft resistance distribution (unit shaft resistance as a function of depth) at EOID and at BOR. For each pile, the EOID unit shaft resistance distribution was subtracted from the BOR unit shaft resistance distribution to yield a unit set-up distribution (unit set-up as a function of depth) for the pile's full length. This determination for one of the 12.75-inch-diameter test piles (Test Pile IPS-12-12) is presented in Figures 1a through 1c.

Cumulative set-up as a function of pile toe depth/elevation was determined for each pile by applying the unit set-up distribution to the surface area of the pile, and cumulatively summing set-up magnitude versus depth. These test program unit shaft resistance and cumulative set-up results are ultimate values. Ultimate cumulative set-up for IPS-12-12 is presented in Figure 2. Komurka (2004) details this approach to characterizing set-up.

Required End-Of-Initial-Drive Capacity

For this example, the desired allowable load is 150 tons. Substituting into Eq. 4 yields:

The relationships presented by Eqs. 1 and 6 are illustrated in Figure 2. The sum of

$$Req'd \quad EOID + SetUp \times \frac{2.25}{2.50} = 150 \quad tons \times 2.25 = 337.5 \quad tons \tag{6}$$

EOID capacity plus unadjusted (actual) cumulative set-up equals the ultimate long-term capacity (which in this example is equal to or greater than 337.5 tons). At any pile toe depth/elevation, the required EOID capacity plus the adjusted cumulative set-up equals 337.5 tons.

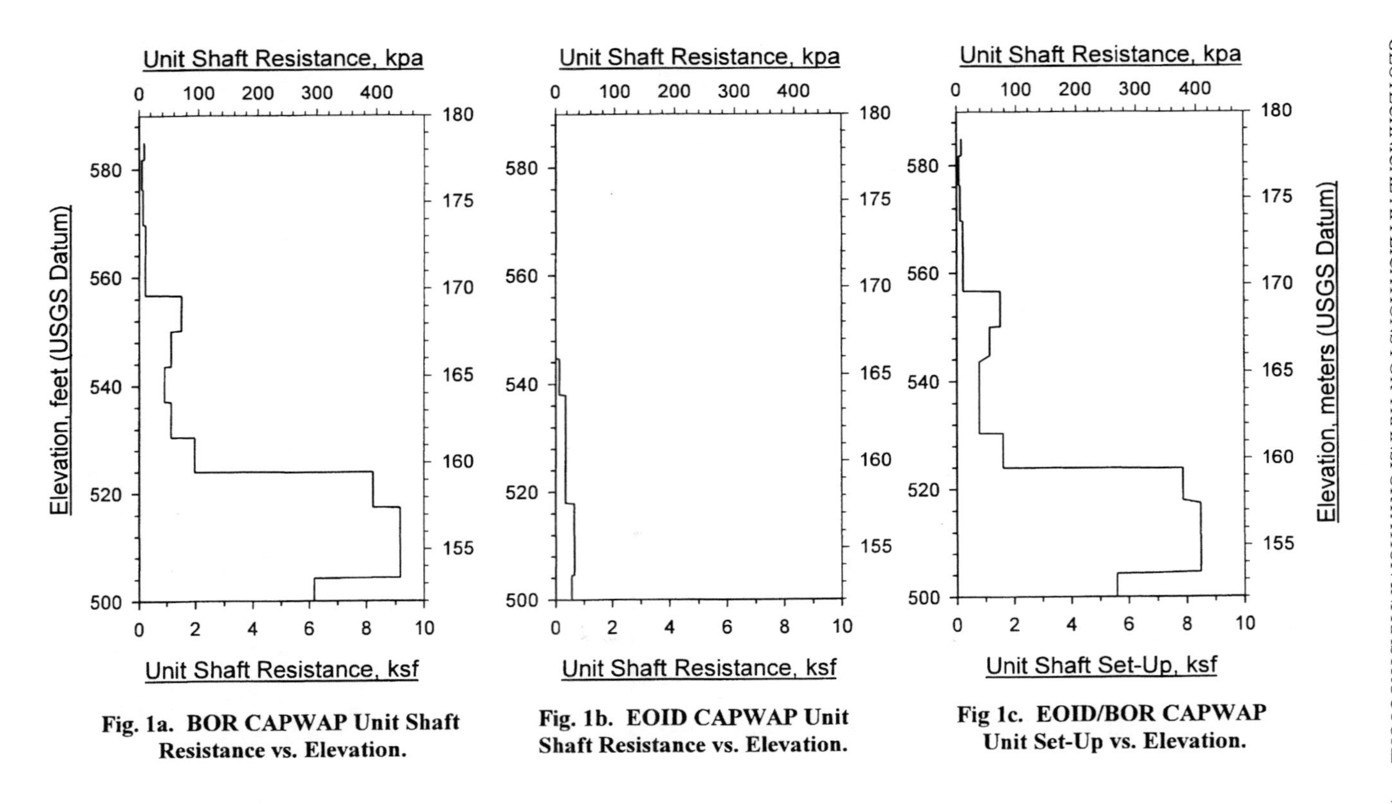

Fig. 1a. BOR CAPWAP Unit Shaft Resistance vs. Elevation.

Fig. 1b. EOID CAPWAP Unit Shaft Resistance vs. Elevation.

Fig 1c. EOID/BOR CAPWAP Unit Set-Up vs. Elevation.

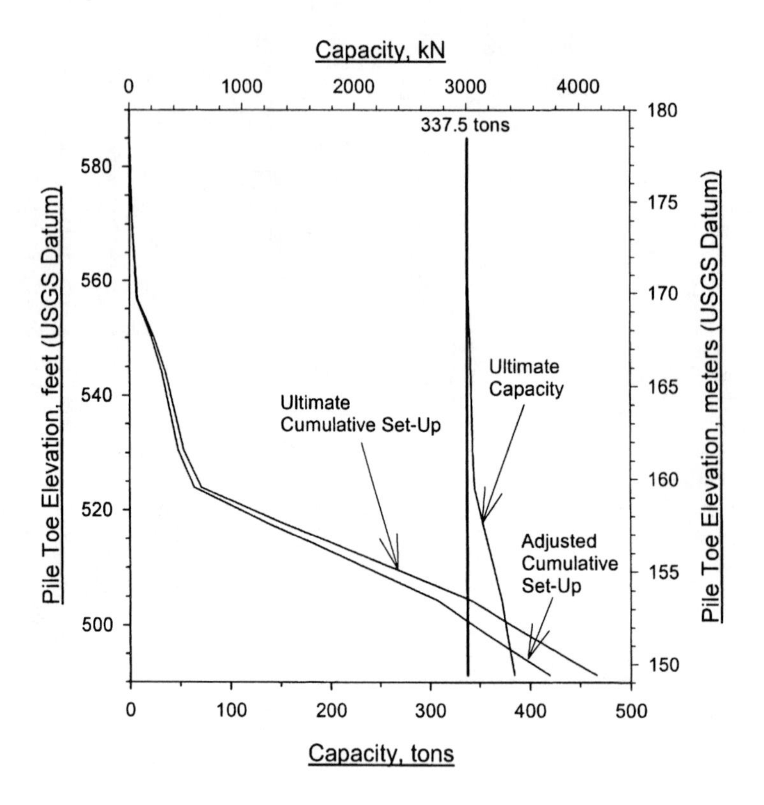

Fig 2. Capacity vs. Pile Toe Elevation.

Rearranging Eq. 6 yields:

$$Req'd \quad EOID = 337.5 \quad tons - SetUp \times \frac{2.25}{2.50} \tag{7}$$

The relationship presented by Eq. 7 for required EOID capacity for a given pile toe elevation is illustrated in Figure 3.

For example, in Figure 3 at pile toe Elevation 513, adjusted cumulative set-up equals 198 tons, requiring 139.5 tons EOID capacity. A review of Figure 3 indicates that since cumulative adjusted set-up increases with pile toe depth, required EOID capacity decreases with depth. This required EOID capacity decrease with depth, from which depth-variable installation criteria can be determined, is illustrated in Figure 4. Depth-variable installation criteria, which account for cumulative set-up increasing with depth in this way, were discussed in Komurka (2004).

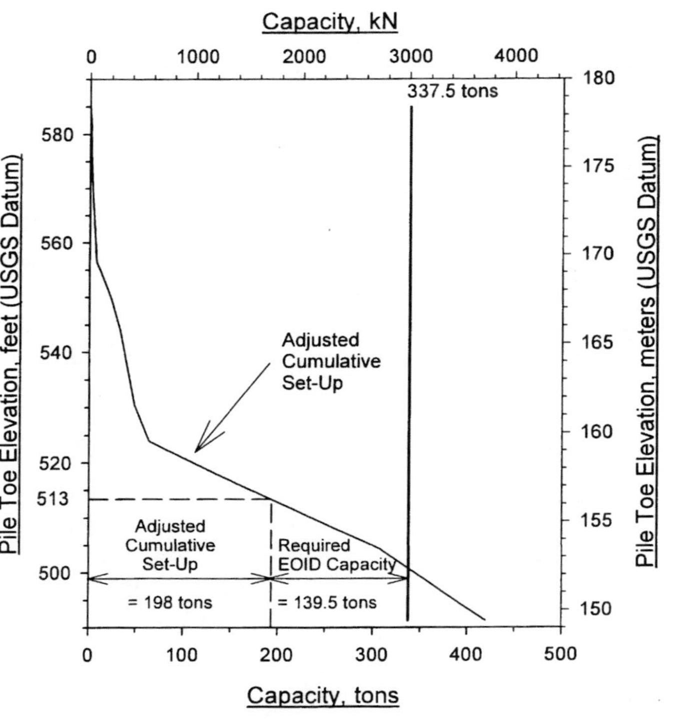

Fig 3. Capacity vs. Pile Toe Elevation.

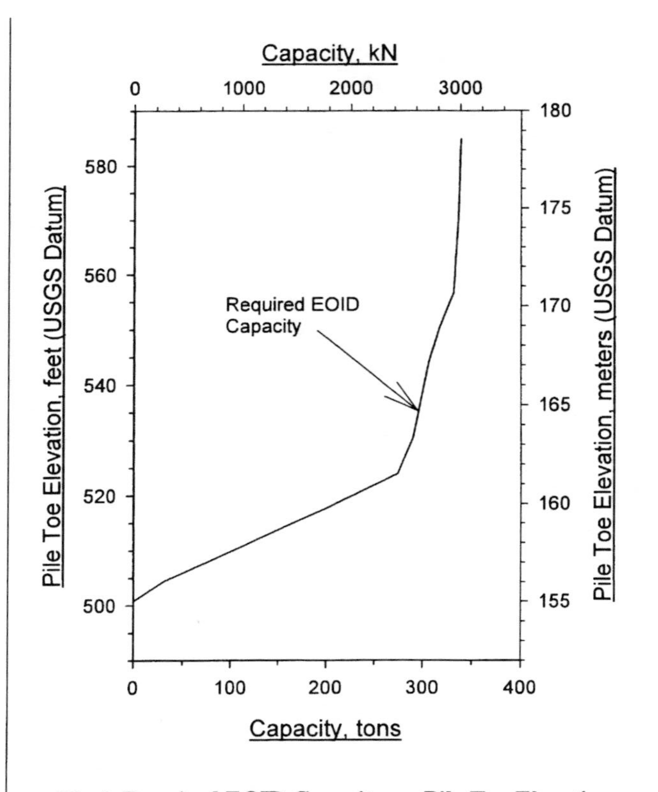

Fig 4. Required EOID Capacity vs. Pile Toe Elevation.

CONCLUSIONS

A driven pile's long-term capacity is often the sum of two components: EOID capacity and set-up. A number of methods may be used to estimate these two capacity components, whether during design or during production driving (i.e., construction control), and the methods may have different degrees of uncertainty. For this reason, and number of others, it may be desirable to apply separate and different safety factors to EOID capacity and set-up.

Safety factor selection may depend on a number of factors, and a number of codes contain selection provisions. Once separate safety factors for EOID capacity and set-up are selected, and a desired allowable load is established, the required EOID capacity to which the piles should be installed can be determined.

APPENDIX I. CONVERSION TO SI UNITS

1 foot (ft) = 0.3048 meters (m)
1 kip per square foot (ksf) = 47.88 kilopascals (kPa)
1 U.S. ton = 8.896 kilonewtons (kN)

REFERENCES

Axelsson, Gary (2002). "A Conceptual Model of Pile Set-up for Driven Piles in Non-Cohesive Soil," *Deep Foundations Congress, Geotechnical Special Publication,* No 116, Volume 1, ASCE, Reston, Va., pp. 64-79.

Bullock, Paul Joseph (1999). "Pile Friction Freeze: A Field and Laboratory Study, Volume 1," Ph.D. Dissertation, University of Florida.

Chellis, R.D. (1961). *Pile Foundations*, Second Edition, McGraw-Hill Book Company, New York, pp. 21-23.

Chow, F.C., Jardine, R.J., Brucy, F., and Nauroy, J.F. (1998). "Effects of Time on Capacity of Pipe Piles in Dense Marine Sand," *Journal of Geotechnical and Geoenvironmental Engineering*, Vol. 124, No. 3, ASCE, pp. 254-264.

Goble, George G., and Rausche, Frank (1970). "Pile Load Test by Impact Driving," *Highway Research Record,* Highway Research Board, No. 333, Washington, DC.

Goble, George G., Likins, Garland E., and Rausche, Frank (1975). "Bearing Capacity of Piles from Dynamic Measurements," Final Report, Department of Civil Engineering, Case Western Reserve University, Cleveland, Ohio.

Goble, George G., Moses, Fred, and Snyder, Richard (1980). "Pile Design and Installation Specification Based on Load-Factor Concept," *Transportation Research Record 794,* National Research Council, Washington, D.C., pp. 42-45.

Guang-Yu, Z. (1988). "Wave Equation Applications for Piles in Soft Ground," *Proc., 3rd International Conference on the Application of Stress-Wave Theory to Piles* (B. H. Fellenius, ed.), Ottawa, Ontario, Canada, pp. 831-836.

Hannigan, Patrick J., Goble, George G., Thendean, Gabriel, Likins, Garland E., and Rausche, Frank (1997). *Design and Construction of Driven Pile Foundations – Volume II*, Federal Highway Administration Report No. FHWA-HI-97-013.

Huang, S. (1988). "Application of Dynamic Measurement on Long H-Pile Driven into Soft Ground in Shanghai," *Proc., 3rd International Conference on the Application of Stress-Wave Theory to Piles* (B. H. Fellenius, ed.), Ottawa, Ontario, Canada, pp. 635-643.

Hussein, Mohamad, Sharp, Mike, and Knight, William (2002). "The Use of Superposition for Evaluating Pile Capacity," *Proc., ASCE GeoInstitute's International Deep Foundation Congress*, Orlando, Florida.

Komurka, Van E., Wagner, Alan B., and Edil, Tuncer B. (2003a). "A Review of Pile Set-Up," *Proc., 51st Annual Geotechnical Engineering Conference*, University of Minnesota, St. Paul, Minnesota, February 21, 2003.

Komurka, Van E., Wagner, Alan B., and Edil, Tuncer B. (2003b). "*Estimating Soil/Pile Set-Up,*" Wisconsin Highway Research Program Contract No. 0092-00-14, Report No. 0305, September 2003.

Komurka, Van E. (2004). "Incorporating Set-Up and Support Cost Distributions into Driven Pile Design," *Current Practices and Future Trends in Deep Foundations,* ASCE/Geo-Institute, Geotechnical Special Publication No. 125, 0-7844-0743-6, 2004, pp. 16-49.

Liang, R., and Nawari, N. (2000). "Evaluation of Resistance Factors for Driven Piles," *New Technological and Design Developments in Deep Foundations,* ASCE/Geo-Institute, Geotechnical Special Publication No. 100, pp. 178-191.

Likins, Garland E., Rausche, Frank, Thendean, Gabriel, and Svinkin, Mark (1996). "CAPWAP Correlation Studies," STRESSWAVE '96 Conference, Orlando, Florida.

Likins, Garland E., (2004). "Pile Testing – Selection and Economy of Safety Factors," *Current Practices and Future Trends in Deep Foundations,* ASCE/Geo-Institute, Geotechnical Special Publication No. 125, 0-7844-0743-6, 2004, pp. 239-252.

Likins, Garland E., Rausche, Frank (2004). "Correlation of CAPWAP with Static Load Tests," Proc., *Seventh International Conference on the Application of Stress-wave Theory to Piles*, Petaling Jaya, Selangor, Malysia, August 9-11, 2004.

Long, James H., Kerrigan, John A., and Wysockey, Michael H. (1999). "Measured Time Effects for Axial Capacity of Driven Piling," *Transportation Research Record 1663,* Paper No. 99-1183, pp. 8-15.

Long, James H. (2002). "Resistance Factors for Driven Piling Developed from Load-Test Databases," *Deep Foundations 2002,* ASCE/Geo-Institute, Geotechnical Special Publication No. 116, 2002, pp. 944-960.

Paikowksy, Samuel G. (2004). "*Load and Resistance Factor Design (LRFD) for Deep Foundations,*" NCHRP Report 507, Transportation Research Board, Washington, DC.

Peck, Ralph B. (1942). "Discussion: Pile Driving Formulas," Proceedings of the American Society of Civil Engineers, Vol. 68, No. 2, pp. 905-909.

Pile Dynamics, Inc. (1998). Pile Driving Analyzer Manual, Model PAK, Cleveland, Ohio.

Rausche, Frank, Goble, George G., and Moses, F. (1972). "Soil Resistance Predictions from Pile Dynamics," *Journal of the Soil Mechanics and Foundations Divi-*

sion, Vol. 98, No. SM9, September, 1972, ASCE, pp. 917-937.

Rausche, Frank, Goble, George G., and Likins, Garland E. (1985). "Dynamic Determination of Pile Capacity," *Journal of Geotechnical Engineering,* Vol. 111, No. 3, ASCE, pp. 367-383.

Rausche, Frank, Hussein, Mohamad, Likins, Garland E., and Thendean, Gabriel (1994). "Static Pile Load-Movement from Dynamic Measurements," *Proc., ASCE Geotechnical Engineering Division's Vertical and Horizontal Deformations of Foundations and Embankments Conference*, College Station, Texas.

Rausche, Frank, Thendean, Gabriel, Abou-matar, H., Likins, Garland E., and Goble, George G. (1996a). *"Determination of Pile Driveability and Capacity from Penetration Tests,"* Final Report, U.S. Department of Transportation, Federal Highway Administration Research Contract DTFH61-91-C-00047.

Rausche, Frank, Richardson, B., and Likins, Garland E. (1996b). "Multiple Blow CAPWAP Analysis of Pile Dynamic Records," STRESSWAVE'96 Conference, Orlando, Florida.

Rausche, Frank, Robinson, B., and Liang, L. (2000). "Automatic Signal Matching with CAPWAP," *Proc., Sixth International Conference on the Application of Stress-Wave Theory to Piles,* São Paulo, Brazil, September 11-13, 2000.

Samson, L., and Authier, J. (1986). "Change in pile capacity with time: Case histories," *Canadian Geotech. Journal,* 23(1), pp. 174-180.

Skov, Rikard, and Denver, Hans (1988). "Time-Dependence of Bearing Capacity of Piles," *Proceedings 3rd International Conference on Application of Stress-Waves to Piles,* pp. 1-10.

Smith, E.A.L. (1960). "Pile Driving Analysis by the Wave Equation," *Journal of the Soil Mechanics and Foundations Division,* Vol. 86, No. 4, ASCE, pp. 35-61.

Soderberg, Lars O. (1961). "Consolidation Theory Applied to Foundation Pile Time Effects," *Géotechnique,* London, Vol. 11, No. 3, pp. 217-225.

Sowers, George F. (1979). *Introductory Soil Mechanics and Foundations*, Fourth Edition, Macmillan Publishing Company, New York, pp. 531-533.

Svinkin, Mark R. (1996). "Setup and Relaxation in Glacial Sand – Discussion," *Journal of Geotechnical Engineering,* Volume 122, No. 4, ASCE, pp. 319-321.

Svinkin, Mark R., and Skov R. (2000). "Set-Up Effect of Cohesive Soils in Pile Capacity," *Proceedings, 6th International Conference on Application of Stress Waves to Piles,* Sao Paulo, Brazil, Balkema, pp. 107-111.

Svinkin, Mark R., (2003). "Discussion — Reliability Concepts In LRFD Design — Or What is A Reasonable Factor Of Safety?," *Deep Foundations*, The Magazine of the Deep Foundations Institute, Winter, 2003, p. 22.

Titi, Hani H., and Wathugala, G. Wije (1999). "Numerical Procedure for Predicting Pile Capacity – Setup/Freeze," *Transportation Research Record 1663,* Paper No. 99-0942, pp. 25-32.

IMPLEMENTATION OF A PILE LOAD TEST PROGRAM IN THE DESIGN PHASE: MARQUETTE INTERCHANGE PROJECT

David A. Rudig[1], P.E., Member, ASCE and Therese E. Koutnik[2], P.E., Member ASCE

ABSTRACT: The $810 Million Marquette Interchange Project is a reconstruction of an interchange located in the heart of downtown Milwaukee, which connects 3 major interstates I-43, I-94, and I-794. The Milwaukee Transportation Partners Geotechnical team for the Marquette Interchange Project was tasked with providing foundation design recommendations at 260 pier locations for 22 mainline and high level ramp bridges. The original interchange was supported by various 420-kN to 800-kN pipe and H-piles. However, for the new interchange, the MTP Geotechnical Team recommended using soil set-up in order to achieve higher capacity piles up to 2225 kN with shorter pile lengths for a greater economy. The recommended foundation types were 1335-kN 324 mm, 1780-kN 356 mm, and 2225-kN 406 mm closed ended pipe piles. As a means to better define the potential cost savings to the project in foundation design, the MTP design team conducted a design phase pile load test program and has projected a project savings for the 2005 to 2008 construction to be as much as $11 Million less the difference between the pile load test program and the original geotechnical investigation cost. MTP determined from a value engineering analysis that if a similar foundation system of a highway project had at least a $2.5 Million construction fee, the project would likely receive a net benefit from performing a design phase pile load test program. This paper not only evaluates the advantages of performing pile load test during the design phase, as opposed to performing verification pile testing during the construction phase, but also addresses the process and implementation of the data obtained in the pile load test program into the final design and specifications.

[1]Principal Engineer, HNTB Corporation, 11414 West Park Place, Suite 300, Milwaukee, WI 53224, drudig@hntb.com
[2]Geotechnical Engineer, HNTB Corporation, 11414 West Park Place, Suite 300, Milwaukee, WI 53224, tkoutnik@hntb.com

INTRODUCTION

The $810 Million Marquette Interchange Project is reconstruction of an interchange located in the heart of downtown Milwaukee, which connects 3 major interstates I-43, I-94, and I-794. As a means to better define the potential cost savings of the project in foundation design, the MTP design team conducted a design phase pile load test program. MTP has projected the project savings for the 2005 to 2008 construction to be as much as $11 Million less the difference between the pile load test program and the original geotechnical investigation cost. This paper not only evaluates the advantages of performing pile load test program during the design phase, as opposed to performing verification pile testing during the construction phase, but also addresses the process and implementation of the data obtained in the pile load test program into the final design and specifications.

STEP 1: JUSTIFICATION OF THE DESIGN PHASE LOAD TEST PROGRAM

Step 1, Justification of the Design Phase Test Program, was one of the most crucial steps in the decision-making process of a project. Results of Step 1 were used to direct the course of the geotechnical program and answered the question that all owners ask: Is the program justified? If it can be demonstrated that a project can gain significant cost savings as a result of completing a design phase load test program; then, it is worth it to proceed.

The first task towards completing Step 1 was to evaluate the feasible foundation options based on soil and groundwater conditions, past performance of adjacent existing foundations, predicted load-capacities of proposed foundations, constructability, and cost of foundation installation and subsequent proof-testing.

For the Marquette Interchange project, Step 1 was completed when MTP prepared the "*Core Investigation Report*" and subsequent Addendum in January 2003. The report addressed 22 bridge structures, supported by about 260 substructure units, located in the Core Area of the Interchange. The report was divided into four major sections that followed the sequence of tasks performed: (1) Review of Existing Subsurface Conditions and Existing Foundations; (2) Development of General Subsurface Conditions; (3) Categorization of Loading Conditions; and (4) Comparison and Evaluation of Feasible Foundation types.

Review of Existing Subsurface Conditions and Foundations

This review of existing conditions was comprised of compiling geotechnical reports and plans for existing structures and utilities located on, or adjacent to, the project from Wisconsin Department of Transportation (WisDOT), Milwaukee County, and Milwaukee Metropolitan Sewerage District. A literature search was also conducted to review local and regional groundwater and geology, topography, soil surveys, and other additional reports of the downtown Milwaukee area soils. This information was summarized to determine the general soil and groundwater conditions; feasible foundation types based on past experience and performance; and what additional subsurface information was needed.

In the Core area, 159 borings were drilled in the 1960s for the original interchange. These borings were located at existing substructure units and ranged from 1 to 55 meters (m) deep with an average depth of 20 m. A total of 31 borings were advanced deeper than 30 m. The depths of these borings were sufficient for the original design which used piles ranging from 9 to 30 m deep with capacities from 420 to 800 kiloNewtons (kN).

To obtain data needed for design of higher capacity foundation elements, new borings had to be extended to greater depths than those drilled in the original investigation. A prototype geotechnical exploration program was developed to obtain additional information. The selected boring locations and number were chosen to provide adequate coverage (approximately one boring per two acres area) throughout all the Core area. They extended deep enough to obtain more reliable information to design any feasible foundation type with the highest expected loading conditions.

The prototype exploration program also determined the extent of the second phase of the geotechnical program that would be required, which was tailored to the foundation type selected. The prototype program consisted of 23 borings that ranged from 37 to 69 m with 6 of the borings reaching bedrock. This program was supplemented by 15 borings that were drilled in 2002 for a previous study for the Marquette project and ranged from 20 to 36.5 m deep. These 38 recent borings were then added to the review of subsurface conditions.

Development of General Subsurface Conditions Profile

Using the information obtained in the existing conditions review and in the 2002-2003 prototype geotechnical programs, the Core project area was divided into four soil sectors, based on similar subsurface profiles.

Sector A was defined as the region that did not contain organics, having a general profile consisting of shallow-depth fill over silty clay or clayey silt (glacial till) that was embedded with erratic lenses of silt and sand (lacustrine). Sector B contained shallow depth unclassified fill over compressible organic silt and clay materials to significant depths underlain by lacustrine and glacial till.

Sector C was similar to Sector B; however, the materials underlying the compressible organic deposits were primarily alluvial. Sector D was similar to Sector A, but consisted of unclassified fill over shallow, soft clay that was not necessarily organic, and was underlain by competent glacial clay till. Because the footing elevations of piers within Sector D were to be set below the soft clays, the piers in Sector D were included in the Sector A discussion.

The top of dolomite bedrock ranged between 41 to 66 m deep, or between 118.7 to 147 meters, MSL. Bedrock was generally shallowest in Sector A-West which was the northwest area of the Core, and then deepened to the south and east.

Shallow groundwater was noted at or above the level of the Menomonee or Milwaukee Rivers, which is about 176.8± meters, MSL. On the boring logs, the majority of water levels were encountered within silty clay or fill strata in Sector A, within the fill or near the top of organic deposits in Sector B, and within the fill masses in Sectors C and D.

Categorization of Loading Conditions

The next task in determining the most feasible foundation type was to categorize the loading conditions of the 22 bridge structures within the Core. Structures were divided by bridge type that had similar forces and moments. Bridge types were determined to be part of either the High Level System Ramp Bridges (Load Set 1), the Low Level Girder Bridges (Load Set 2), and the Menomonee Valley Bridge Widening (Load Set 3). A summary of all load sets are summarized in the Table 1.

TABLE 1. Summary of Forces and Moments for Load Sets 1 to 3

Load Set (1)	Axial Forces (kN) (2)	Shear Forces (kN) (3)	Transverse Moments (kN-m) (4)	Longitudinal Moments (kN-m) (5)
1	13,345-17,79	867-1,157	1,356-16,270	3,390-20,337
2	3,558-17,793	NA	0-3,932	0-6,779
3	1,334-31,582	0-778	542-20,337	54-11,117

Load Set 1, the System Ramps and I-43 bridges, produced high moments in both the transverse and longitudinal directions. These high moments were a result of using curved alignments, single shaft columns, and higher pier heights of these bridges.

Load Set 2, the low level girder bridges, included those along the existing I-794 eastbound and westbound mainlines. The low level girder bridges produced lower axial forces and moments than those in Load Set 1. The lower forces were the result of shorter span lengths, lower pier heights, and multi-column piers.

Load Set 3, the Menomonee Valley Bridge widening, was considered a girder bridge. It had lower axial forces and moments similar to Load Set 2, where the bridge widths were smaller. It had higher axial forces but similar moments to Load Set 1 where the bridges were wider and higher than the System Level Ramps. The loads calculated for the widening assumed that the proposed footings would not be connected nor transfer any loads to the existing foundations.

Comparison and Evaluation of Feasible Foundation Types

By dividing the soil conditions into 3 Sectors and categorizing the bridges into 3 load sets, MTP created a matrix which regarded nine different combinations. The matrix provided the vehicle to optimize the foundation type for each substructure unit, as these two factors (soil sector and load set) influenced the foundation performance, constructability, and cost.

Shallow foundations or footings were considered, along with deep foundations consisting of either precast concrete piles, H-piles, closed ended pipe piles, open ended pipe piles, or drilled shafts. Shallow foundations or footings were not suitable to support the high loading conditions and also satisfy the overall external stability and serviceability requirements of these structures.

Precast concrete piles were eliminated as a feasible alternative. They appear to be best suited for areas where the final toe elevations are expected to be relatively consistent. The variable composition and strength of the soil deposits within the Core required variable foundation lengths. Field splices and/or cutoffs would be numerous, and special care during driving to prevent pile damage. In addition, this pile type is not commonly driven in Wisconsin.

With precast piles and shallow foundations excluded, the four remaining foundation types were evaluated based on general advantages and disadvantages, performance, constructability, and cost for the three loading sets (Load Sets 1 to 3) and three soil sectors (Sectors A, B, and C).

The performance of each foundation type was determined by estimating (1) the average allowable lateral capacity, (2) the required unit length, (3) the group array, and (4) the group settlement. The allowable lateral capacity was defined as the lateral resistance of a foundation corresponding to 25.4 mm deflection. This value was averaged between two rows for pile and shaft groups. The foundation lengths were then determined using static methods and a pre-selected allowable capacity. For pile foundations, the pre-selected allowable capacities are summarized in Table 2. For the drilled shafts, MTP considered a group of multiple 1.2-m shafts, a single 2.4-m shaft, and a single 3.6-m (12-foot) shaft. The allowable capacities were based on the loading conditions per load set.

The allowable capacities of the high capacity closed ended concrete filled pipe piles with diameters of 324, 356, and 406 mm (otherwise noted as high capacity piles), incorporated set-up or soil freeze within their estimated capacities. For the report, the capacity was estimated by using the effective stress method and from previous experience of soil set-up in the Milwaukee area. Based on other studies in the downtown Milwaukee, soil set-up was anticipated to account for up to 55 percent of ultimate pile capacity. If this general foundation type was selected, the lengths would have to be verified with a pile load test program.

TABLE 2. Pre-selected Allowable Capacity for Pile Foundations

Pile Type (1)	Pile Size (2)	Selected Allowable Pile Capacity (kN) (3)
H-piles	250x62	489
	310x79	578
Closed End Pipe Piles	273mm	489
	305mm	578
	324mm	1335
	356mm	1512
	406mm	1780
Open End Pipes	762mm	3559
	914mm	4448
	1067mm	5338

A foundation's constructability considered available right-of-way, utilities, vibration effects, noise generation during construction, and speed of installation. Construction time for a substructure unit was determined, based on common local experience of installation rates and the estimated number of piles or shafts per group.

The average cost for a substructure unit was calculated, including materials and labor for construction of the foundation units and the connecting cap, any additional testing, backfilling, and construction monitoring. In order to compare the foundation types equally and to calculate the potential foundation costs for the overall project, the cost of each foundation type in each load set-soil sector combination was divided by the average axial load and expressed as dollars per supported ton.

In general, MTP determined that high capacity concrete-filled closed-end piles with diameters in the range of 324, 356, and 406 mm were the most economical foundation choices with single 8-foot drilled shafts as a close second alternative for all soil sectors and load sets.

High capacity piles outweighed the other options because of the economy of using time-dependent soil set-up in the design capacity. Selection of the high capacity piles assumed that a design phase pile load test would be conducted in order to better define the accuracy of the estimated lengths of the piles. Costs ranged from \$5.8 to \$8.9 per supported kN for the high capacity piles, with the next alternative 4 to 47% higher. The second choice was generally 20 to 47% higher than the high capacity piles, where the majority of the substructure units were located.

The potential cost saving by using the high capacity piles instead of conventional piling was estimated to be as much as $11 Million less the difference between the pile load test program from the original geotechnical program, about $300,000.

The $11 Million savings could only be achieved if a pile load test program was conducted during the design phase of the project and not during construction. At the time this program was proposed, WisDOT had performed very few such programs on any prior projects in the state. Federal Highway Administration (FHWA) encouraged the design-phase load test approach.

There was discussion calling for deferring all the load testing to the construction phase. However, the potential for cost overruns, change orders, and time delays due to redesign could be substantial, considering the number of substructure units on this project. Though all these factors were important, avoiding delays was the most critical, as the net time of public travel inconvenience was considered paramount.

Also, if only a construction phase pile load test program was conducted, a more conservative factor of safety and thus a lower allowable capacity would have been assigned during the design phase. This would have increased the pile lengths, number of piles, and cap sizes. Therefore, Step 1 justified a design phase pile load test for the Marquette project and projected that it was well worth considering.

STEP 2: DEVELOPMENT OF THE PILE LOAD TEST PROGRAM

The second step of the process was to develop and design an appropriate pile load test program and the second phase of the geotechnical investigation.

A second phase geotechnical investigation could include additional borings, more laboratory and/or in-situ testing, static or dynamic axial load tests, lateral load tests, uplift load tests, and setup evaluations on indicator piles. The number, location, and type of testing are all dependent on the purpose, the size of the project, the variability of the subsoil, the controlling loading conditions, the physical distance between substructure units, and the comfort levels of the owner and engineer. Generally, the greater number of borings and testing increases the accuracy of results, which in turn, decreases the risk and number of unknowns on a project. Continually increasing the testing can produce diminishing returns; however, having an inadequate number of tests generates more conservative assumptions and thus more costly results. The engineer and owner together must determine the most reasonable level of testing to produce the greatest rate of return without compromising the data required to design the foundations.

From a value engineering analysis performed, MTP determined if a similar project had at least $2.5 million foundation construction fee, the project is likely to have a net benefit from performing a design-phase load test program. Such an analysis is dependent on the subsurface conditions and the loads on foundation units.

On the Marquette Interchange project, the purpose of the pile load test program included: (1) To confirm suitability and constructability of piles with 1335-kN or greater allowable loads using static compression load tests and indicator piles; (2) To quantify soil set-up magnitude, distribution, and time rate across the site; (3) To quantify and confirm lateral load performance and p-y curve assumptions via lateral load tests, and to extrapolate this information across the site by characterizing the subsurface materials with in-situ pressuremeter testing; (4) To quantify installation conditions for high-capacity piles in a summary report, thereby allowing foundation contractors to limit contingency costs in their foundation bids. Implementation of pile test program has shown to minimize construction costs and provide contractors with definitive data on expected situations regarding the installation of driven piles.

The level of testing was a greatly discussed topic, and as shown below in Table 3 and Figure 1, the greatest benefit to cost ratio was between Program D, E and F.

TABLE 3. Cost / Benefit of Varying Geotechnical Investigations

Program (1)	A (2)	B (3)	C (4)	D (5)	E (6)	F (7)
No. Static Load Tests	0	4	5	6	8	11
No. Indicator Piles	0	24	36	46	55	75
Boring Program & Lab Testing Cost (millions)	2.1	1.8	1.5	0.8	0.8	0.6
Pile Load Test Program Cost (millions)	0.0	1.2	1.5	1.6	2.0	3.0
Total Program Cost (millions)	2.1	3.0	3.0	2.4	2.8	3.6
Potential Savings (millions)	0.0	3.0	5.0	8.0	11.0	13.0
Benefit/Cost Ratio	0.0	1.0	1.7	3.3	3.9	3.6

The final geotechnical investigation included 65 additional deep borings ranging from 36.5 to 69 m (120 to 226 feet) deep; various laboratory testing; 9 pressuremeter locations with up to 5 tests per pile test-setup location. This varied testing approach differed from the conventional FHWA and WisDOT approach, which was to have one soil boring drilled to refusal at every pier location, 260 borings. The savings in the boring and lab testing program selected (Program D) from the typical program (Program A) exceeded $700,000.

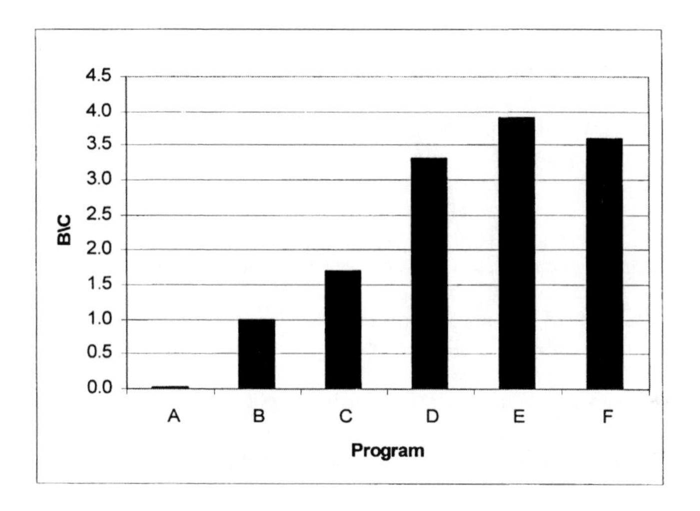

FIG. 1. Benefit / Cost Ratio of Varying Geotechnical Investigations

The final load test program consisted of 6 static axial load tests, 3 lateral load tests, and 42 indicator pile sites spread throughout the core site area. For the 88 piles that were driven, 3 timed restrikes were conducted per pile, as well as vibration monitoring at select locations. Therefore, the projected savings for the final load test program was estimated to be between about $8.5 and $11 Million.

The static, reaction, and indicator piles were either 324-mm outside diameters (OD) x 9.53-mm wall, 356-mm OD x 12.7-mm wall, or 406-mm OD x 12.7-mm wall. All piles were ASTM A-252 Grade 3 steel with minimum yield strength of 310 kilopascals (kPa).

The static axial load tests were loaded until failure, or up to 5338 kN. Testing was performed no earlier than 28 days after pile installation. Three timed restrikes were scheduled at (1) 18 to 24 hours; (2) about 10 days after installation; and (3) after minimum of 28 days after installation. Piles were installed with a Delmag D46-32 or Junttan HHK-10A hammer. Lateral load tests were conducted at the static load test locations by jacking two reaction piles away from each other to a maximum of 400 kN. Inclinometers recorded load-deflection behavior versus depth.

The pile locations were placed in areas where the greater number of production piles would be installed, but yet included the general soil profiles. The percentage of bridge piers within each soil sector was estimated to be 35% in Sector A, 59% in Sector B, and 6% in Sector C. The 6 static load tests and 42 indicator piles were divided as 38% in Sector A, 52% in Sector B, and 10% in Sector C. The total cost of the design phase pile load test program was $2.4 Million, including engineering, of which 54% was the pile installation.

This load test program was supplemented by the prototype geotechnical investigation consisting of 38 borings as previously described in Step 1 and by the Core retaining wall geotechnical program. The investigation for Core walls included 15 borings, 9 cone penetration tests, 4 pressuremeter test locations, 2 dilatometers, and 5 monitoring wells. The retaining wall geotechnical program costs were not included in the above table, as they were performed primarily for another purpose.

The total number of borings performed in the Core (118 borings) did not equal the final 260 total number of substructure pier units, for several reasons. First, the existing borings drilled in the 1960s aided in determining the relative subsurface profile. Secondly, a better method to estimate pile lengths and capacities was to actually drive the same types of piles that will be used in the foundation design. The 42 total pile test locations gave the design team and prospective contractor bidders insight into the issues that may occur during construction, as well as to better predict foundation costs. Thirdly, for environmental investigations of the upper-zone unclassified fills, Geoprobes about 4.6 to 7.6 m deep were advanced at every one of the 260 substructure units to determine if obstructions were present, and to identify any impacted fill materials that might be encountered during the excavation of the pile caps.

MTP proposed 148 fewer borings in the final program, but provided more useful information in terms of data, value, and potential cost savings by providing a more diverse geotechnical program that included the pile load testing program.

STEP 3: CONTRACTING AND IMPLEMENTATION OF THE PILE LOAD TEST PROGRAM

Performing a load test program during the design phase presented two contracting options: (1) Have the load tests performed and evaluated by a specialty contractor and its engineering subconsultants; that is, contracting directly with WisDOT through a formal bidding process, using specifications developed by the design team, or (2) Have the load tests performed by a specialty pile driving contractor and engineering subconsultants retained under the design team's contract with WisDOT; then having the design team select the specialty contractor and consultants by an informal quality-based selection (QBS) process.

For the sake of maintaining an aggressive preliminary engineering design schedule and submittals, WisDOT waived the formal bidding process and agreed that the second contracting option be followed, because the design team would be using this data to complete final design and plan documents. WisDOT directed that the specialty contractors submit competitive quotations and a detailed workplan based on a scope of work developed by the design team. The design team and engineering subconsultants were to provide all instrumentation, conduct all tests and evaluate all results for the design team.

The design team solicited proposals and letters of interest from three invited contractors who would install and restrike all static load test piles and all indicator piles. They were to provide all test frames and equipment. An informal QBS selection process, following interviews with two most responsive proposers, selected Marquette Constructors.

Criteria for contractor selection included past experience performing axial and lateral pile load tests to plunging failure; knowledge of, and prior use of, timed restrikes to estimate gains in capacity due to setup effects; ability to provide experienced field and administrative staff with knowledge of the instrumentation to be used in these tests; ability to perform work on the schedule developed by the design team; and prior experience with installing high-capacity piles not driven to refusal.

At the same time, the design team interviewed and engaged specialty engineering firms for providing instrumentation, conducting static, lateral, and pressuremeter tests with their own testing equipment, to assist the design team in the evaluation of test results. These firms were pre-approved by WisDOT, and in this case, were recommended for this project by either WisDOT or FHWA. GRL Engineers Inc, WKG^2, and STS Consultants respectively, were selected to assist the design team.

During the selection and contracting process, the design team developed a sequence of field operations with the specialty contractor and developed a comprehensive schedule for the delivery of the following reports:

The Pile Load Test Program Data Report was a compilation of all the data generated during the Pile Load Test Program, which included driving records, dynamic monitoring results, end-of-drive and set-up distributions versus elevation, cumulative set-up versus elevation, static axial compression load tests, pile head load deflection curves from lateral load and static load tests, set-up versus time plots, and vibration monitoring data.

The Summary of Pile Load Test Program for Bidding Contractors was intended to be an executive summary report of and included portions of the Pile Load Data Report.

The Geotechnical Foundation Design Reports (GFDR) were intended to be used by structural designers in order to complete the design for each bridge. A total of 22 reports were generated. Each report included a summary of the borings, existing foundation data, and pile load test program data relevant to each bridge. Recommendations included (1) Applicable soil profiles and p-y curves for lateral load requirements; (2) Assigned allowable load versus elevation curves at each substructure unit; (3) Allowable uplift versus elevation curves; (4) Group spacing requirements; (5) Additional provisions related to vibrations; and (6) Removal of existing foundations and/or coordination with other bridges or retaining walls.

The Construction Implementation Report was intended to be used by the owner's representative, who also would perform verification pile load testing during construction. The report outlined the methodology and the process of how to use the data in the Pile Test Program Data Report in order to assess production pile installations.

Precondition surveys of nearby structures, layout of indicator piles and test pile groups; obtaining access permits and utility clearances preceded installation of any piles. These steps covered about 7 weeks. Mobilizing, installing and restriking the static, reaction, and indicator piles required an additional 13 weeks. The axial and lateral load tests were performed concurrent with the installation of indicator piles, over a period of 8 weeks. Evaluation of all test results covered approximately 14 weeks. Providing the four reports as noted in the contract required approximately 9 weeks. Total duration of the program if performed consecutively would have taken about 43 weeks.

STEP 4 – FINAL DESIGN AND CONSTRUCTION

The intention of the pile load test program in the design phase was to provide specific recommendations for optimum pile diameters, wall thicknesses, and lengths to be used in various sectors under varying load conditions. This information was used to develop final designs and interim construction cost estimates.

Verification that certain pile sizes and significant capacity gains from soil set-up could be realized, had a direct effect on construction cost estimates. It was critical to be able to predict and control the foundations budget for the project. Information on driveability criteria, vibration damage potential and access issues valuable to the future bidders, was to be provided in the deliverables in the previous step. It was useful to the design team in developing performance specifications.

The final design of the bridge foundations included determining the length of the piles for specific axial loads, checking the group action for lateral and uplift, and reducing axial capacity for downdrag effects if necessary. MTP foundation engineers used the GFDR content to complete the final design. Lengths were estimated by using the allowable capacity versus depth curves that were assigned to each substructure unit. These curves were determined by dividing the load test results (end-of-drive plus set-up) by a factor of safety. Lateral deflection of the group was determined by using a computer program with user-input custom p-y curves generated for varying soil types across the project. Uplift was checked by using the allowable uplift (soil set-up divided by a safety factor) versus depth curves assigned to each substructure unit. Axial capacity was reduced if the relative settlement of the soils adjacent to the piles was assumed to be greater than 12.7 mm. Typically, estimated settlements were due to adjacent fill walls or embankments.

The final design of the foundations was illustrated on project plans and specifications. The MTP plans typically included a foundation plan sheet that stated the loading conditions, pile types, allowable capacity, and estimated length per pier. The lengths were estimated to the nearest 0.61 m, and the allowable capacities were estimated to the nearest 45 kN. The allowable capacities and pile sizes were selected as follows: up to 1335 kN for 324-mm OD x 9.53-mm wall; 1780 kN for 356-mm OD x 12.7-mm wall; and up to 2225 kN for 406-mm OD x 12.7mm wall. The allowable capacity per pile could vary between each pier but not within the same pier.

The above recommendations were made in order to minimize construction errors and to simplify the design. The majority of the bridges determined that the 356 mm pile up to 1780 kN allowable capacity was generally the most economical. Minimum tip elevations were only shown on plans if required for lateral capacity or downdrag conditions. The design end-of-drive capacity was not included because verification or dynamic pile testing will be conducted during construction.

The technical specifications were part of the Special provisions, which addressed (1) dynamic pile installation and restrike testing and (2) delivery and installation of the high capacity piling.

Because indicator piles were not driven at each substructure unit during the design phase, verification testing of test piles or dynamic pile testing would still be necessary to establish the final driving criteria at each pier. The Marquette Interchange project requires a minimum of 10 percent (rounded down to the nearest integer), or two tested piles per pier, whichever is greater. The dynamic pile installation and restrike load testing are to be performed by the owner's representative, who will have three major assigned tasks; (1) Perform wave equation analyses for acceptance of the contractor's pile driving equipment; (2) Perform dynamic installation and restrike testing on test piles; and (3) Generate final driving criteria to be used by inspectors for acceptance of production piles.

The Marquette Interchange project required the contractor to submit all pile driving equipment for approval prior to any driving. The required number of hammer blows indicated by the wave equation analysis at the ultimate pile capacity (equal to 1.7 times the allowable capacity) must range between 30 and 120 blows per 0.305 m. The pile stresses must also not have exceeded 90 percent of the steel pile yield stress. Grade 3 (minimum of 310 MPa) steel and 27.58 MPa concrete was specified for the high capacity pile materials.

The driving criteria for hammer acceptance was a debated topic between the design team and the owner. Initially, an acceptable hammer was intended to drive the piles to an ultimate capacity of twice the allowable load. However, the wave equation analysis for the first structure to be built demonstrated that the contractor's equipment would not have been acceptable at each pier. The criteria was then reduced to 1.7 because that was the minimum required for mobilization of the test piles.

The next step was to complete the dynamic installation and restrike load testing. The Construction Implementation Report was generated by MTP in order to provide the methodology and step-by-step field instructions for the dynamic testing to any engineering firm selected as the owner's representative. On the Marquette Interchange project, GRL Engineers, Inc. has been awarded this role.

For the dynamic testing, the test piles at each pier were to be driven to the specified length on the plans and prior to any production piles. At the specified length, a CAPWAP analysis was to be performed to verify if the required end-of-drive capacity as designed was achieved. If not, the pile would then be driven deeper until the end-of-drive criteria was satisfied. Once the end-of-drive criteria was accepted, the test piles were to be re-struck 18 to 72 hours after installation to verify that the design set-up capacity was being achieved.

The set-up could be calculated if the pile was mobilized. If the pile was not mobilized, it was assumed that a factor of safety of 1.7 was reached. If the pile was mobilized, the measured set-up would be checked against the predicted or design set-up value. If this was greater or equal to the predicted set-up, then the construction load testing was considered to be complete. If this was not satisfied, then multiple restrikes were required.

After the dynamic testing was complete for all the test piles within a substructure unit, a final driving criteria, or minimum penetration resistance criteria, was to be generated by the owner's representative and given to the inspector for acceptance of the production piles. The final driving criterion would include a minimum specified length and hammer blow count chart (blow count versus pile embedment depth) beyond the minimum pile length per substructure unit.

In set-up conditions in general, the hammer blow count on the chart decreases with increasing pile length because as the pile lengthens, the setup capacity increases while the required end-of-drive capacity decreases for the same design capacity. The contractor then would drive the piles to the length specified in the final driving criteria. If the hammer blow count would be more than the one shown on the hammer blow count chart, the production piles would be acceptable. If not, the contactor would continue driving until this occurred.

CONCLUSIONS

Before the load testing program was proposed, MTP calculated that the high-capacity pile foundation alternative appeared to have the lowest overall cost and would produce the greatest potential project savings, over conventional piling, if allowable capacities greater than 1335 kN per pile could be proved in a design phase load test program.

MTP determined from a value engineering analysis that if a similar foundation system of a highway project had at least a $2.5 Million construction fee, the project would likely receive a net benefit from performing a design phase pile load test program. The advantages of conducting a design phase load test program included using soil set-up in the design, decreasing the number of piles and the overall pile lengths, reducing pile cap sizes, using lower factors of safety, and accelerating the construction schedule. Prospective bidders would be given valuable insight into foundation-related construction issues, thus reducing their overall risk and producing better foundation bids.

MTP proposed fewer numbers of borings than the total substructure units, but then provided much more in terms of data, value, and potential cost savings. The final geotechnical program included 118 borings, 6 Static load tests, 3 lateral load tests, and 42 indicator pile locations. Having the contractor for the load test program retained by the design team proved to be successful resulting from a more aggressive design schedule. The expected savings over conventional piling was estimated to be between $8.5 and $11 Million less the difference between the pile load testing program and the original geotechnical investigation, about $300,000. This proved to be one of the largest single cost-saving measures for the entire project.

REFERENCES

Milwaukee Transportation Partners (2003). "Core Investigation Report, Marquette Interchange Project, Project ID 1060-05-03." Milwaukee Transportation Partners.

Milwaukee Transportation Partners (2003). "Addendum to Core Investigation Report, Marquette Interchange Project, Project ID 1060-05-03." Milwaukee Transportation Partners.

EVALUATION OF GEOSYNTHETICS USE FOR PAVEMENT SUBGRADE RESTRAINT AND WORKING PLATFORM CONSTRUCTION

Erol Tutumluer[1], Member, ASCE and Jayhyun Kwon[2]

ABSTRACT: Subgrade restraint design is the use of a geosynthetic placed at the subgrade/subbase or subgrade/base interface to increase the bearing capacity or the support of construction equipment over a weak or soft subgrade. The conventional design criteria of unpaved roads require providing adequate base course or aggregate cover material to prevent bearing capacity type failure. Geogrids and high-strength woven geotextiles can increase bearing capacity of a pavement structure. Increasing the bearing capacity of subgrade soils can reduce required base course or treatment thickness. Several design methodologies exists for the use of geosynthetics in subgrade restraint for pavement construction. This paper evaluates currently available design approaches and design tools including related proprietary software. Comparative analysis results for different design approaches are also presented in this paper.

INTRODUCTION

Geotextiles and geogrids are the most popular type geosynthetics used in the road construction industry. Both geogrids and high-strength woven geotextiles can perform as tensile reinforcement for aggregate base courses. Geogrids and high strength geotextiles can also be used to increase stability and improve performance of weak road foundation soils by providing subgrade restraint during construction. Subgrade restraint design is the use of a geosynthetic placed at the subgrade/subbase or subgrade/base interface to increase the bearing capacity or the support of construction equipment over a weak or soft subgrade. Adding a geosynthetic layer can increase bearing capacity of a pavement structure by forcing the potential bearing capacity surface to develop along alternate, higher shear strength surfaces. The lateral restraint and/or membrane tension effects may also contribute to load carrying capacity. Geosynthetics can also be used over weak subgrade soils to provide a working platform for construction equipment. This paper is intended to provide

[1] Associate Professor, University of Illinois, Department of Civil and Environmental Engineering, 205 N. Mathews, Urbana, IL 61801, tutumlue@uiuc.edu
[2] Graduate Research Assistant, University of Illinois, Department of Civil and Environmental Engineering, 205 N. Mathews, Urbana, IL 61801, jaykwon@uiuc.edu

present knowledge on the increased stability and improved performance of weak subgrades by providing subgrade restraint during construction. Currently available design approaches and design tools including related proprietary software are evaluated in this paper.

CURRENT DESIGN METHODS AND TOOLS FOR SUBGRADE RESTRAINT

The two most widely used geosynthetic subgrade restraint design procedures are the Steward et al. (1977) and the Giroud and Noiray (1981) procedures, which are essentially based on theoretical models. In addition, a number of new design methodologies and software programs, as design tools, have recently been developed to evaluate the potential use of fabrics in the design and construction of geosynthetic reinforced working platforms over soft subgrade soils. Next, these design methods and tools will be described in detail.

U.S. Army Corps of Engineers design of geosynthetic reinforced unpaved roads

The design methodology proposed by Steward et al. (1977) was adopted as the current design procedure in the U.S. Army Corp of Engineers in the Technical Manual 5-818-8 for geotextile reinforced low-volume unpaved roads. The use of the design methodology is limited to the design of low-volume unpaved roads for the approximately 1,000 80-kN (18-kip) equivalent single axle loads (ESALs) causing less than 50-mm (2-inches) of rutting. The essential steps involved in designing for subgrade restraint recommended in TM 5-818-8 are given as follows:

Step 1. Evaluate subgrade soil strength by determining the soil's cohesion parameter, c.
Step 2. Determine the design wheel loading and estimate traffic.
Step 3. Select a bearing capacity factor, Nc.
- Steward et al. (1977) proposed a value of 2.8 for unreinforced roads and 5.0 for geotextile-reinforced roads.
Step 4. Compute the subgrade bearing capacity (c*Nc).
Step 5. Determine required aggregate depth with and without a geotextile from the appropriate design curve (single, dual, or dual tandem load configuration).

Results of previous full-scale test sections were reviewed in a recent study to validate the empirical bearing capacity factors, Nc, used for unreinforced and geotextile-reinforced base materials (Tingle and Webster, 2003). In 1995, the airfield pavement division of the Army Corps Engineers constructed four full-scale test sections over a very soft high-plasticity (CH) clay subgrade (California bearing ratio or CBR of 1 percent). The four sections constructed included one control section and three geosynthetic-reinforced sections. All the test sections were designed to have similar level of deterioration for a design traffic level of 2,000 military truck passes. Five hundred passes of 186-kN (41.9-kip) gross weight 5-ton military truck were used to simulate the construction traffic.

The empirical bearing capacity factors, Nc, were determined from the test results (Tingle and Webster, 2003). The experimental bearing capacity factors, Nc, of 2.6,

3.6, and 5.8 were obtained for the unreinforced, geotextile reinforced, and the geogrid-geotextile reinforced sections, respectively. The researchers concluded that the design bearing capacity factor of 2.8 for the unreinforced section was appropriate based upon limited full-scale data. Since the experimental bearing capacity factor for the geotextile reinforced section, 3.6, was less than the existing Corps of Engineers design value of 5.0, additional testing was recommended to modify the existing specifications. On the other hand, the geogrid-reinforced section had a backcalculated Nc value of 5.8 greater than the 5.0.

Giroud-Han (2004) design method

A new design methodology was recently developed by Giroud and Han (2004a-b) for the subgrade restraint application of geogrids and geotextiles manufactured by Tensar Earth Technologies, Inc. As an add-on to the previous design methodology by Giroud and Noiray (1981), the new Giroud-Han method incorporates the strength/ modulus of the base material, variations of the stress distribution angles through the base course, and the aperture stability modulus of the geosynthetic reinforcement. The method is theoretically based and empirically calibrated.

The Giroud-Han method was calibrated with the results of a specially developed research program, which incorporated a large number of large-scale, cyclic plate load tests using geogrid reinforcement, conducted at the North Carolina State University (Gabr, 2001). Data were gathered from the research study, which considered the pressure induced subgrade and surface deformations as a function of the number of load cycles for multiple combinations of reinforcement and base thicknesses. These data were then used to estimate the pressure distribution angle and quantify the effects of base reinforcement and thickness on both initial stress distribution angle and changes to the angle with continued load applications. Following calibration, the Giroud-Han method was verified using results of other research and field data (Giroud and Han, 2004a-b).

Based on the new Giroud-Han method, the SpectraPave2™ software was developed by Tensar Earth Technologies, Inc. (http://www.tensarcorp.com) to support analysis and design of geosynthetic reinforced flexible pavements and unpaved roads The software utilizes the stiffness and durability of Tensar biaxial geogrids to reinforce soils or embankment fill and enhance the performance of the underlying subgrade or aggregate base and subbase courses. The SpectraPave2™ software features four modules including: (1) project information, (2) subgrade improvement, (3) base course reinforcement, and (4) subgrade improvement cost analysis. The subgrade improvement module incorporates the Giroud-Han method and indicates the required thickness for an unreinforced subbase/base and for a subbase/base reinforced with Tensar geogrids. The base course reinforcement module features serviceability and design applications for both unreinforced and reinforced pavement sections. Note that unreinforced alternatives are based on the 1993 AASHTO Pavement Design Guide and reinforced alternatives are based on empirical traffic benefit ration methodologies.

According to the SpectraPave2™ analysis results, the use of both Tensar geotextiles and geogrids can effectively reduce the aggregate thickness requirements when compared to the unreinforced section results. Geogrids with higher tensile strength and high aperture stability moduli were found to give overall higher geosynthetic stiffnesses and hence work better than geotextiles. For example, the use of the BX1200 geogrid reduced a 406-mm (16-in.) aggregate cover requirement down to only 127-mm (5-in.), more than one-third reduction, for a subgrade soil CBR of 2.5.

Propex's RACE (Roadways and Civil Engineering) design methodology

The RACE design software was developed by Propex Fabrics and Fibers Company to design new flexible pavements or unpaved roads and surfaces with geotextiles (http://www.geotextile.com/). The main features of the RACE design software include: (1) unpaved road design module, (2) flexible pavement design module, and (3) pavement rehabilitation module. Flexible pavement design module was based on the 1993 AASHTO pavement design methodology. The unpaved road design module presents the required thickness with or without the use of geotextile reinforcement so that the user can compare the cost-benefit data for using geosynthetics. The input variables include loading conditions, base course aggregate properties (CBR, aggregate angularity, maximum aggregate size), and subgrade soil conditions (subgrade CBR and subgrade moisture).

After reviewing previous studies conducted by several researchers, Steward et al. (1977), Barenberg (1980), Giroud and Noiray (1981), and Haliburton et al. (1981), the RACE program was based on the finding that the placement of a geotextile beneath an aggregate section increases the permissible stress on a subgrade by a factor of 1.64 to 2.0. The authors of the RACE design software therefore recommended using an average design factor of 1.8. In other words, it is assumed that the subgrade soil will behave the same under a given stress level without a geotextile as at 1.8 times that stress level when a geotextile is placed beneath the aggregate section. Further, it was also recommended to use an allowable stress level of 2.8*cohesion for the unreinforced section and 5.0*cohesion when a geotextile is used as reinforcement. These stress levels should result in minimal rutting, less than 50-mm (2-in.), after approximately 1,000 ESALs or equivalent single axle loads.

In the RACE program, it is assumed that each geotextile type produces approximately the same reduction in the subgrade stress. Therefore, the most important criteria used in the selection of the appropriate geotextile are the fabric construction and long-term survivability. The RACE design software recommends geotextile selection based on the AASHTO specification M288, the National Guideline Specifications for Geotextiles, which is based on geotextile survivability. Further, the RACE design literature quoted states that previous research using geogrids in stabilization also showed approximately the same benefit as geotextiles. The difference is that the geogrids are higher in cost and their open fabric structure does not provide the important separation function unless a geotextile is placed beneath them. In summary, the RACE and SpectraPave2™ programs evaluate geogrids quite differently for the subgrade restraint application and the granular base cover requirements.

Design methodology for thin asphalt pavements on soft soils in the Netherlands

A structural evaluation and design procedure was recently proposed for thin asphalt roads by the Dutch funding agency CROW (van Gurp and van Leest, 2002). The researchers claim that traditional design procedures for highway pavements are not fully applicable to low-volume roads, because of the low traffic intensities, and comparatively high axle loads. The low-volume roads have poor load-carrying capacity. The base course, therefore, generally provides the load-carrying capacity for the low-volume asphalt pavements. The new design methodology proposed takes into account cracking and deformation aspects of the base course. Required thickness of the base course can be calculated using heir proposed equation as follows:

$$h_d = \frac{125.7\log(N_{constr}) + 496.52\log(P) - 294.14RD_{constr} - 2412.42}{f_{undr}^{0.63}} \quad (1)$$

where, h_d = desired total base thickness in construction stage (m);
N_{constr} = number of axle loads in construction stage;
P = average axle load in construction stage (N);
RD_{constr} = allowable rut depth at surface in construction stage (m);
f_{undr} = undrained shear strength of subgrade (Pa) = (20 or 30)*CBR.

A factor of 20 is used to estimate CBR, when the ground water level is high, and a factor of 30 is used when the ground water table is deeper than 0.5 m below the bottom of the base course.

A decision support tool was also provided for assessing the effects of geosynthetics in unbound base courses (van Gurp and van Leest, 2002). Based on the subgrade characteristics and the required base course thickness, the designer can determine the type of geosynthetic applicable from Table 1. As indicated, minimal benefit can be gained in stiff subgrade (CBR > 8) conditions.

The reinforcement benefit can mostly be quantified as an extension of pavement life or a reduction in the required base course thickness. According to this CROW design methodology; pavement life extension is limited to a factor of 4 to prevent unrealistic substantial increases in the pavement life. The ability of geosynthetic reinforcement for reducing base thickness can be deduced from Figure 1 for a variety of geosynthetic types. The maximum reduction in base thickness is limited to 150 mm (approximately 6 in.) and the minimum layer thickness of the reinforced base is set to be 150 mm (approximately 6 in.).

FHWA design methodology

The methodology currently used by the FHWA was developed by Steward et al. (1977) for the design of low volume roads with geotextile and aggregate placed over low bearing capacity subgrade soils. The FHWA design was presented in the form of design charts/procedures and is independent of geosynthetic product material properties. The procedure also includes selecting a bearing capacity factor. The stress level acting on the subgrade can be expressed in terms of the bearing capacity factor. In the FHWA design procedure, selecting the bearing capacity factor is based on the tolerable rut depth and the number of axle passes. The required aggregate thickness can then be determined from the design chart with the obtained bearing capacity factor. The FHWA design guidelines are, in general, limited to a subgrade unconfined compressive strength of less than 90-kPa (approximately a CBR of 3).

TABLE 1. Qualitative overview of potential reinforcement properties of geosynthetics (van Gurp and van Leest, 2002)

Design criterion			Geosynthetics					
Road type	Subgrade class	Base thickness (mm)	Geotextile, GT		Geogrid, GG		GG-GT, composite	
			Non-woven	woven	extruded	woven	Open graded material	Well graded material
Paved	Weak	150-300	●	●	●	□	●	○
	CBR<3	>300	●	●	◆	◆	◆	○
	Moderate	150-300	○	◆	●	□	●	○
	3<CBR<8	>300	○	○	◆	□	●	○
	Stiff	150-300	○	○	◆	□	□	
	CBR>8	>300	○	○	○	○	○	○

Note: Asphalt cover thickness is 80-mm (3.15-in.) in all cases

●: Normally applicable, ◆: Sometimes applicable,

□: Insufficient information available, ○: Normally not applicable

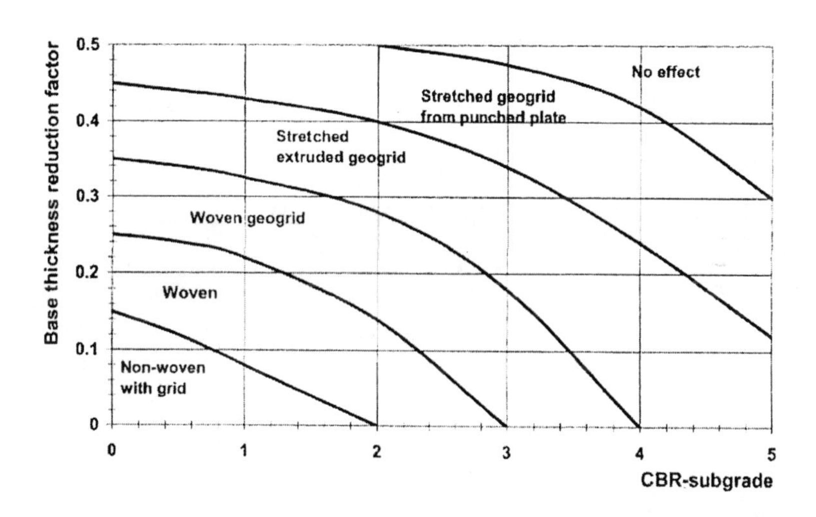

FIG. 1. Reinforcing effect of geosynthetics on base thickness reduction (van Gurp and van Leest, 2002)

As a summary, Table 2 lists from all the studies the empirical bearing capacity factors, Nc, recommended for use with the unreinforced and reinforced pavement sections.

TABLE 2. Summary of recommended bearing capacity, Nc, factors

Design by	Date	Traffic	Rutting	Nc Factor Reinforced	Nc Factor Unreinforced	Comments
Barenberg	1975			6	3.3	
Steward et al.	1977	> 1000 ESALs	Defined as little	5	2.8	Unpaved road
	1998			5	2.8	Used by FHWA for temporary roads and construction platform
	2003	< 1000 ESALs	< 50 mm (2in.)	5	2.8	Used by the US Army Corps of Engineers (US Army COE)
Tingle and Webster	2003	2000 trucks and 500 military trucks				Validation study US Army COE
		< 2000 passes	< 76 mm (3 in.)	3.6	2.6-2.8	Geotextile (GT)
		< 2000 passes	< 76 mm (3 in.)	5.8	2.6-2.8	Geogird (GG)
TENSAR's SpectraPave2	2004	< 1000 ESALs	< 76 mm (3 in.)	5.14 (GT) 5.71 (GG)	2.8	Geotextile (GT) and geogrid (GG)
Propex's RACE	2002	< 1000 ESALs	< 50 mm (2 in.)	5	2.8	Geotextile (GT) and geogrid (GG)

COMPARISONS OF DESIGN REQUIREMENTS FOR UNPAVED ROADS

Comparisons are made in this section among the Illinois Department of Transportation (IDOT) currently used thickness requirements and the SpectraPave2™, RACE, CROW, and Federal Highway Administration (FHWA) design methodologies for unreinforced base course or aggregate cover requirements. First the SpectraPave2™, RACE, and CROW results are evaluated based on similar rut depth criteria used in these programs for subgrade restraint design. Later, the current IDOT thickness requirements are evaluated based on the results of the geosynthetic reinforced design approaches and the thicknesses obtained using the IDOT Subgrade Stability Manual (1982). The design conditions used in all the comparative analyses are summarized in Table 3.

The IDOT CBR based thickness design chart (IDOT Subgrade Stability Manual, 1982) is based on a 50-kN (12-kip) Equivalent Single Wheel Load and 500 number of axle passes. To realistically compare results, the SpectraPave2™, RACE, and CROW analyses were also conducted for the same 50-kN (12-kip) single wheel load, and a tire pressure of 550 kPa (80 psi). The minimum number of axle passes in the RACE program is 1,000. Therefore, the SpectraPave2™, RACE, and CROW analyses were also made for 1,000 axle passes. Unpaved temporary roads usually allow some rutting to occur and are subjected to fewer than 10,000 load applications during their service lives.

TABLE 3. Summary of design conditions for SpectraPave2™, RACE, CROW, and FHWA design methods

	SpectraPave 2	RACE	CROW	FHWA
Loading				
Axle load	100 kN	100 kN	100 kN	100 kN
Wheel Load	50 kN	50 kN	50 kN	50 kN
Tire pressure	550 kPa	550 kPa	550 kPa	550 kPa
Number of passes	1,000	1,000	1,000	1,000
HMA surface	-	-	76 mm	-
Aggregate				
Base CBR	70	Crushed medium hard rock (60 - 80)	-	-
Aggregate angularity	-	Subangular	-	-
Lift thickness	-	150 mm (minimum)	-	-
Maximum aggregate size	-	38 mm		-
Subgrade				
Subgrade moisture	-	Subgrade moisture is not likely impact construction	Ground water level is deeper than 0.5 m below the bottom of the base course	-
Soil condition	-	Fine grained soil with smooth surface	-	-

The required base course thicknesses obtained from the IDOT, SpectraPave2™, RACE, CROW, and the FHWA design methods for unpaved roads are compared in Tables 4 and 5 and in Figures 2 and 3. It is important to note that all design procedures have different failure criteria as mentioned in the previous paragraph. Therefore, the differences in the predicted aggregate thicknesses are due to both different rut depth conditions and also how each method estimates the subgrade stress. No rut information is given in Table 4 for the IDOT analysis due to the fact that sinkage equation used for the IDOT chart is not comparable to the bearing capacity theory related rut depths used in the other analysis programs. Nevertheless, the IDOT thickness requirements are in good agreement with the FHWA, CROW, and RACE results with 50 mm (2.0 in.) of rut depth. There are negligible differences between the aggregate thicknesses obtained from the CROW design and the RACE program and both results for the unreinforced case fall in between the SpectraPave2™ results. With geotextile reinforcement, the RACE thicknesses for a 50-mm (2.0-in.) rut condition were also in between the SpectraPave2™ thickness predictions for the 38-mm (1.5-in.) and 76-mm (3.0-in.) rut conditions.

Comparing the results presented for with and without geosynthetics cases, the aggregate thicknesses can be considerably reduced using geosynthetics. Approximately the same aggregate thicknesses are required according to the FHWA design method and the SpectraPave2™ analysis with geotextile using a tolerable rut depth of 76-mm (3.0-in.) (see Table 5 and Figure 3). The SpectraPave2™ design methodology is based on the approach adopted by Giroud-Han (2004a-b) for geosynthetic–reinforced unpaved roads.

TABLE 4. Results obtained using IDOT, SpectraPave2™, RACE, CROW, and FHWA design methods for unpaved roads (without geosynthetics)

Subgrade CBR	Required base thickness of *unreinforced* unpaved road (mm)					
	SpectraPave2™		RACE	CROW	FHWA	IDOT
	Rut = 38 mm	Rut = 76 mm	Rut = 50 mm	Rut = 50 mm	Rut = 50 mm	-
1.5	660	389	485	523	413	442
2	533	305	401	437	347	389
2.5	465	254	330	378	280	350
3	447	229	274	338	240	318

TABLE 5. Results obtained using SpectraPave2™, RACE, and FHWA design methods for unpaved roads (with geosynthetics)

Subgrade CBR	Required base thickness of *reinforced* unpaved road (mm)							
	SpectraPave2 (Rut = 38mm)			SpectraPave2 (Rut = 76 mm)			RACE	FHWA
	Geotextile	BX1100	BX1200	Geotextile	BX1100	BX1200	Rut = 50 mm	Rut = 50 mm
1.5	475	401	251	264	213	119	310	267
2	378	315	183	193	147	102	224	200
2.5	312	257	145	137	102	102	152	147
3	292	234	127	102	102	102	152	107

A thinner aggregate layer is required with geotextile by the FHWA design method when compared to the RACE program – 107 mm (4.2-in.) thick compared with 150 mm (6.0-in.) – for a subgrade CBR of 3. The base aggregate used in FHWA method had a minimum CBR of 80, whereas the RACE analysis results are based on base course CBR values ranging from 60 to 80 and the SpectraPave2™ analysis considered a CBR of 70. The aggregate quality, considered with the CBR input, significantly influences the stress distribution and performance of the pavement. As a result, the CBR input for the aggregate cover will result in different benefits for the geosynthetics reinforced unpaved roads.

According to the SpectraPave2™ results, the aggregate thicknesses are typically reduced in accordance with the type of geosynthetics used in the subgrade restraint application. For the greatest benefit, a high tensile strength geogrid, such as the BX1200, offers the maximum reduction in the aggregate thickness required (see Figure 3). Whereas, a woven geotextile is not as effective as geogrids, which is indicated in Table 5 and Figure 3 by the lower thickness reductions. Tingle and Webster (2003) also obtained similar results in the recent US Army Corps of Engineers study.

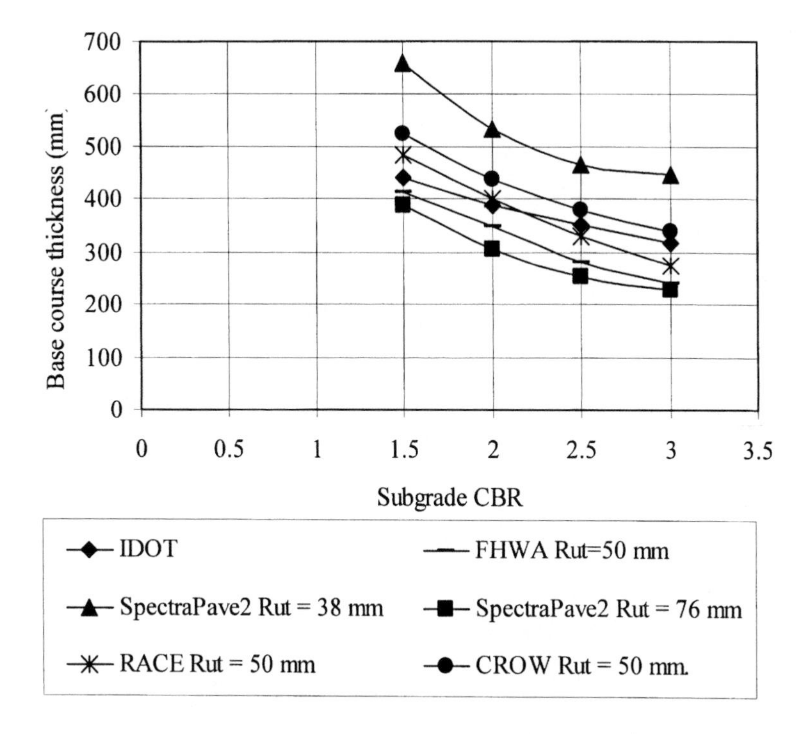

FIG. 2. Results obtained using IDOT, SpectraPave2™, RACE, CROW, and FHWA design methods for unpaved roads (without geosynthetics)

SUMMARY AND CONCLUSIONS

The use of geosynthetics in unpaved roads and flexible pavement sections can lead to considerable improvements in pavement performance. Subgrade restraint design is the use of a geosynthetic placed at the subgrade/subbase or subgrade/base interface to increase the bearing capacity or the support of construction equipment over a weak or soft subgrade. A recent survey conducted among state highway agencies indicated that geosynthetics were more likely being used in the US for subgrade restraint rather than base reinforcement (Christopher et al., 2001). After a thorough review of the current and ongoing geosynthetics research and the currently available design tools and methodologies presented in this paper, there is no doubt that the use of geotextiles and geogrids is beneficial in the subgrade restraint design.

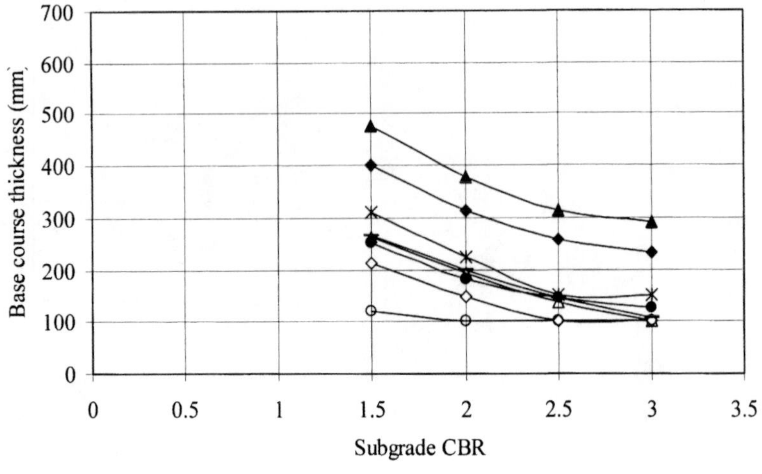

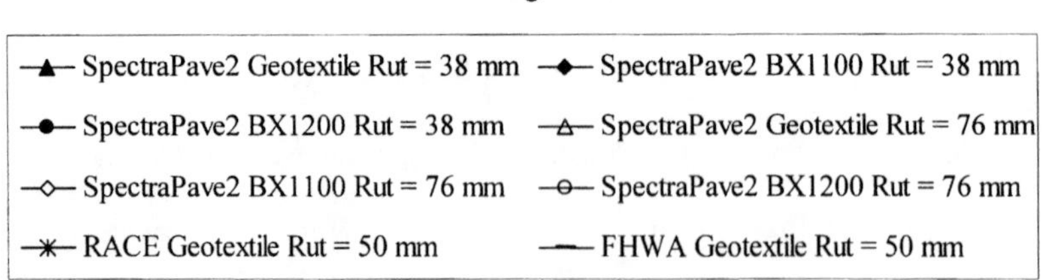

FIG. 3. Results obtained using SpectraPave2™, RACE, and FHWA design methods for unpaved roads (with geosynthetics)

Considerable decreases in the required base thicknesses were indicated from the SpectraPave2™, RACE, and FHWA analyses for California bearing ratio or CBR values less than 6. Note that some of these thickness reductions predicted may not be conservative for very low CBR values less than 1.5. Therefore, the option or decision of decreasing substantially the base course thickness should be taken with caution. A maximum thickness reduction of 150 mm (6 in.) was for example suggested by the CROW methodology for a more conservative use of geosynthetics for subgrade restraint. This is due to the fact that the design methodologies are probably not very well defined for those low CBR values and thus might be inadequate to directly apply to soft subgrade conditions. Nevertheless, some rather thick aggregate cover may very well justify reducing thicknesses even greater than 150 mm (6 in.) under certain circumstances. Further, the potential cost-benefits of using geosynthetics for reducing required aggregate thickness over the subgrade soil should always be evaluated and possibly verified from field effectiveness or demonstration studies.

REFERENCES

Barenberg, E.J. (1980). "Design Procedures for Soil-Fabric-Aggregate Systems with Mirafi 500X Fabric," Final Report, University of Illinois at Urbana-Champaign, UILU, EN-802019.

Christopher, B. R., Berg, R. R, and Perkins, S. W., (2001). "Geosynthetic Reinforcement in Roadway Sections," NCHRP Synthesis for NCHRP Project 20-7, Task 112: Final Report, 119p.

Gabr, M. (2001). "Cyclic Plate Loading Tests on Geogrid Reinforced Roads," Research Report on Tensar Earth Technologies, *Inc.*, NC State University, 43 p.

Giroud, J.P. and Noiray, L. (1981). "Geotextile Reinforced Unpaved Road Design," *Journal of the Geotechnical Engineering Division*, ASCE, Vol. 107, No GT9, 1233-1254.

Giroud, J.P. and Han, J. (2004a). "Design Method for Geogrid-Reinforced Unpaved Roads I: Development of Design Method," *Journal of Geotechnical and Geoenvironmental Engineering*, ASCE, Vol. 130, No. 8, August, 775-786.

Giroud, J.P. and Han, J. (2004b). "Design Method for Geogrid-Reinforced Unpaved Roads II: Calibration and Applications," *Journal of Geotechnical and Geoenvironmental Engineering*, ASCE, Vol. 130, No. 8, August, 787-797.

Haliburton, A.T., Lawmaster, J. D., and King, J.K. (1981). "Potential Use of Geotechnical Fabric in Airfield Runway Design," Contract No. AFOSR 79-00871, Air Force Office of Scientific Research, United States Air Force, Bolling AFB, Washington, D.C.

Illinois Department of Transportation (1982) "Subgrade Stability Manual," IDOT Bureau of Materials and Physical Research, Springfield, Illinois, March, 66 p.

Steward, J.E., Williamson, R. and Mohney, J. (1977). "Guidelines for Use of Fabrics in Construction and Maintenance of Low-Volume Roads," USDA, Forest Service Report PB-276 972, Portland, OR, 172 p.

Tingle, J.S. and Webster, S.L. (2003). "Corps of Engineers Design of Geosynthetic-Reinforced Unpaved Roads," In *Transportation Research Record 1849*, TRB, National Research Council, Washington DC, 193-201.

van Gurp, C.A.P.M. and van Leest, A.J. (2002). "Thin Asphalt Pavements on Soft Soil," In 9th International Conference on Asphalt Pavements, ISAP, Copenhagen, 1-18.

ALTERNATIVE TESTING TECHNIQUES FOR MODULUS OF PAVEMENT BASES AND SUBGRADES

Auckpath Sawangsuriya[1], Peter J. Bosscher[2], P.E., Member, ASCE, Tuncer B. Edil[2], P.E., Member, ASCE

ABSTACT: The importance of stiffness measurements has gained increased recognition in geotechnical applications in pavements. Two alternative testing techniques: bender elements and soil stiffness gauge (SSG) have been recently adopted as they show some potential and promising means of monitoring the stiffness and/or modulus of pavement materials. Since each technique has its own range of stress and strain levels, the relationship between the elastic moduli and nonlinear behavior exhibited by soils at large strains is required so that the measured modulus can be adjusted or corrected to a modulus corresponding to the desired strain levels. This paper presents the implications of these testing techniques in stiffness and/or modulus assessment of pavement bases and subgrades. To adjust the modulus measured in these materials, the desired strain amplitudes must be known. The strains incurred in the pavement base and subgrade layers that are subjected to the typical traffic loadings are summarized from a number of studies including finite-element analyses, large-scale model experiments, and in-situ test sections. The typical range of strain amplitudes imposed by the bender elements and the SSG is compared with those incurred in the pavement base and subgrade layers to evaluate their suitability in the assessment of pavement layer stiffness and/or modulus. Finally, some comments on the practical implications of these techniques to monitor the pavement layer stiffness and/or modulus are provided.

[1] Graduate Research Assistant, Department of Civil and Environmental Engineering, University of Wisconsin-Madison, Madison, WI 53706, sawangsuriya@wisc.edu
[2] Professor, Department of Civil and Environmental Engineering and Geological Engineering Program, University of Wisconsin-Madison, Madison, WI 53706.

INTRODUCTION

For proper design of a pavement, the stress and strain conditions within the pavement structure due to traffic loading must be determined. The states of stress and strain are not only a function of the traffic loading but also of the moduli of the various pavement layers whereas these moduli are in turn a function of the stress state. Stiffness and/or modulus are the engineering properties that are needed for the evaluation of the long-term pavement performance. In pavement engineering, the resilient modulus is a laboratory measure of the elastic modulus of soil under various states of stress common within pavement layers and has been commonly adopted as the important property for characterizing pavement materials especially the design of flexible pavements. The test generally consists of a number of loading steps, where the specimen is subjected to different confining pressures and deviator stresses at each step as well as a number of loading cycles. In addition to the resilient modulus test, the alternative testing techniques such as the soil stiffness gauge and the bender elements can be chosen as the supplementary tests for characterizing pavement materials. The objective of this paper is to describe the implications of these alternate methods for obtaining moduli for pavement bases and subgrades.

ALTERNATIVE TESTING TECHNIQUES FOR SOIL MODULUS

In additional to the resilient modulus test commonly used in pavement design, two currently available modulus tests are considered as alternative testing techniques. The first method is to employ bender elements to measure the travel time of an elastic wave propagating through the soil. This method is robust in that it can be combined with a variety of geotechnical laboratory tests and also shows a great potential for future use in monitoring the stiffness at small-strains by utilizing the characteristic of elastic wave propagation in different media. The second method is to utilize the soil stiffness gauge (SSG), which has been developed to measure the in-place surficial soil stiffness by means of electro-mechanical vibration. A brief description of each method is given below.

Bender Elements

An elastic wave propagation technique that utilizes two-layer piezoceramic bender elements as source and receiver provides a means of measuring the shear wave velocity and the corresponding small-strain shear modulus. The bender element test has become increasingly popular in a variety of geotechnical laboratory applications (Dyvik and Madshus 1985; Thomann and Hryciw 1990; Souto et al. 1994; Fam and Santamarina 1995; Zeng and Ni 1998; Fioravante and Capoferri 2001; Pennington et al. 2001; Mancuso et al. 2002). The transmitting bender element produces a shear wave (S-wave) which propagates through the soil when it is excited by an applied voltage signal. This S-wave impinges on the receiving bender element, causing it to bend, which in turn produces a very small voltage signal. Fig. 1 illustrates typical input and output signals from the transmitting and receiving bender elements.

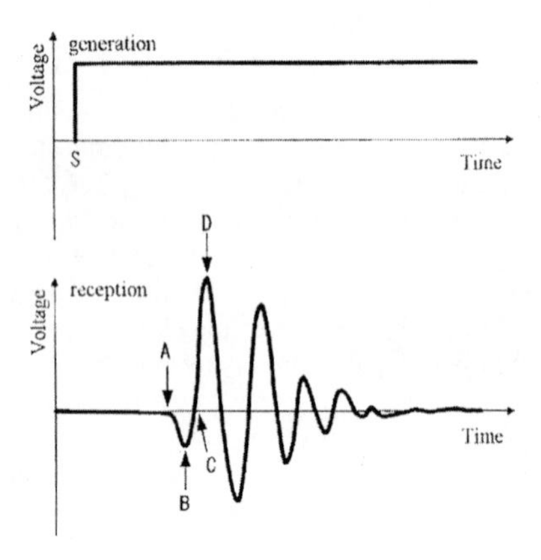

FIG. 1. Typical input and output signals from the transmitting and receiving bender elements

In general, the signals may be different from those in Fig. 1, possibly due to the stiffness of the soil, the boundary conditions, the test apparatus, the degree of fixity of the bender element into the platen or housing, and the size of the bender element and specimen. By measuring the travel time of the S-wave and the tip-to-tip distance between transmitting and receiving bender elements, the S-wave velocity of the soil is obtained. The small-strain shear modulus (G) can be calculated according to elastic theory using the measured S-wave velocity (v_s) and total density of the soil (ρ) with the relationship $G = \rho v_s^2$.

Soil Stiffness Gauge

A recently developed equipment called the soil stiffness gauge (SSG) is a portable, non-nuclear, and non-destructive testing device that provides an alternative means of rapidly and directly assessing in-place surficial stiffness and/or modulus of soils at small strains. Unlike other modulus tests, the operation of the SSG is relatively simple and does not require a skilled operator. Moreover, this device has built-in capability to make computations in order to acquire the stiffness and/or modulus of test materials. Further information and operation of the SSG is provided by Humboldt (1999). The test is conducted in accordance with ASTM D 6758, Standard Test Method for Measuring Stiffness and Apparent Modulus of Soil and Soil-Aggregate In-Place by an Electro-Mechanical Method. Because of its rapid and direct measurement, the SSG appears to have real potential as a supplementary non-nuclear method for earthwork quality control (Edil and Sawangsuriya 2005). The modulus from the SSG appears to correspond to a strain amplitude larger than the strain amplitude of the seismic test, even though the SSG induces a strain amplitude comparable to that of a seismic test (Sawangsuriya et al. 2003). It looks as if the modulus reported by the SSG has been internally reduced by a factor (Sawangsuriya et al. 2003; 2004) possibly to correspond to the resilient modulus. However, this is not disclosed by the manufacturer.

RESULTS AND ANALYSIS OF MODULUS TESTS

Modulus from Bender Elements

A bender element test was performed on a specimen 36 mm in diameter and 78 mm high subjected to a range of confining pressures. The base and top cap were modified to include the transmitting and receiving bender elements respectively inserted in the soil specimen. In order to verify the bender element test, the test was initially conducted on dry Ottawa 20-30 sand, for which published modulus data are available. Note also that in this study, the travel time of the S-wave is determined on the basis of time-domain analysis in such a way that the first arrival in output signal is determined manually. The first arrivals corresponding to points B and C (Fig. 1) were chosen for the determination of the travel time in order to avoid the near-field effect in a triaxial specimen due to wave reflection from the cap boundaries (Arulnathan et al. 1998). This near-field effect may mask the arrival of the S-wave when the distance between the source and receiver (d) is in the range ¼ to 4 wavelengths (λ), which can be estimated from $\lambda = v_s/f$ where v_s is the S-wave velocity and f is the mean frequency of the received signal (Mancuso and Vinale 1988). In this study, the ratio d/λ ranged from 1.6 to 2.5.

Fig. 2 shows the plot of shear modulus obtained from the bender element (BE) tests as a function of confining pressure. Results from the bender element tests were also compared with those from the resonant column (RC) tests which were conducted at two shear strain levels, $5\text{x}10^{-3}$ and $1\text{x}10^{-2}$%. Note that the resonant column tests were conducted on an identically prepared specimen in terms of dry density and method of preparation. The shear modulus of this sand was also computed using the empirical equations given by Hardin and Black (1968) and Seed and Idriss (1970). The shear moduli obtained from the bender element test, the resonant column test, and the empirical equations are compared as shown in Fig. 2. Results indicated good agreement with those suggested by Hardin and Black (1968) and Seed and Idriss (1970).

Another series of bender element tests were conducted on dry medium sand specimens (36 mm in diameter and 78 mm high). Five methods of specimen preparation: (1) scooping, (2) tamping, (3) rodding, (4) vibrating, and (5) pluviating were used. Details of these methods are described in Sawangsuriya et al. (2004). Fig. 3 shows the plot of shear modulus of these sands prepared by these five methods as a function of confining pressure. The shear moduli of sands as measured by the bender element (BE) tests follow the general dependency of modulus on confining stress. Results obtained from the resonant column (RC) tests at two shear strain amplitudes (i.e., $5\text{x}10^{-3}$ and $1\text{x}10^{-2}$%) are also shown in Fig. 3 (results of replicate testing). The modulus-confining stress relationships obtained from the BE tests compare well with those from the RC tests. In general, the moduli from the BE tests are greater than those from the RC test at shear strain amplitude of $1\text{x}10^{-2}$%.

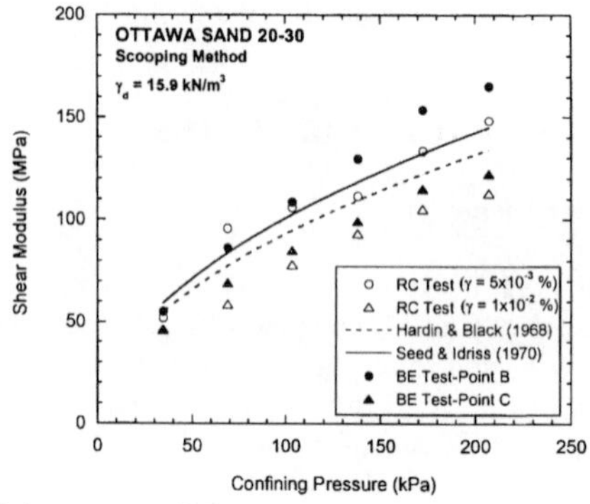

FIG. 2. Shear modulus vs. confining pressure relationship for dry Ottawa 20-30 sand

Modulus from the SSG

Dry medium sand was placed in a 0.3-m radius cylindrical cardboard mold to a height of 530-mm following the five methods described above for the SSG and seismic tests (Sawangsuriya et al. 2004). The SSG test was conducted at the center of the top of the specimens. For the rigid ring-shaped foot of the SSG resting on sand, the measured stiffness of sand (K_{SSG}) from the SSG can be converted to the shear modulus (G) as follows (assuming a linear-elastic, homogeneous, and isotropic infinite half-space which is appropriated for small strains induced by the SSG):

$$G = \frac{K_{SSG}(1-\nu)}{3.54R} \tag{1}$$

where ν is the Poisson's ratio of the soil and R is outside radius of the ring-shaped foot of the SSG.

Comparison of Measured Modulus with Maximum Modulus

The modulus from the bender element and the SSG tests are compared with the maximum modulus (denoted by G_o or G_{max}), which establishes a benchmark value of modulus for deformation problems (Burland 1989). This maximum modulus defines the starting point of a modulus degradation curve (the variation of the modulus with log strain amplitude) and is useful for defining the initial modulus of an empirical stress-strain curve for nonlinear models of soil behavior (Hardin and Drnevich 1972; Jardine et al. 1986; Jardine and Potts 1988; Tatsuoka et al. 1993).

In this study, G_o for the sand specimens prepared by the five methods of specimen preparation was obtained using a pulse echo test. The pulse echo test was performed after the SSG test. A compressional wave (P-wave) was generated by an impulse source at the top of the sand specimens using a hand-held hammer and an aluminum

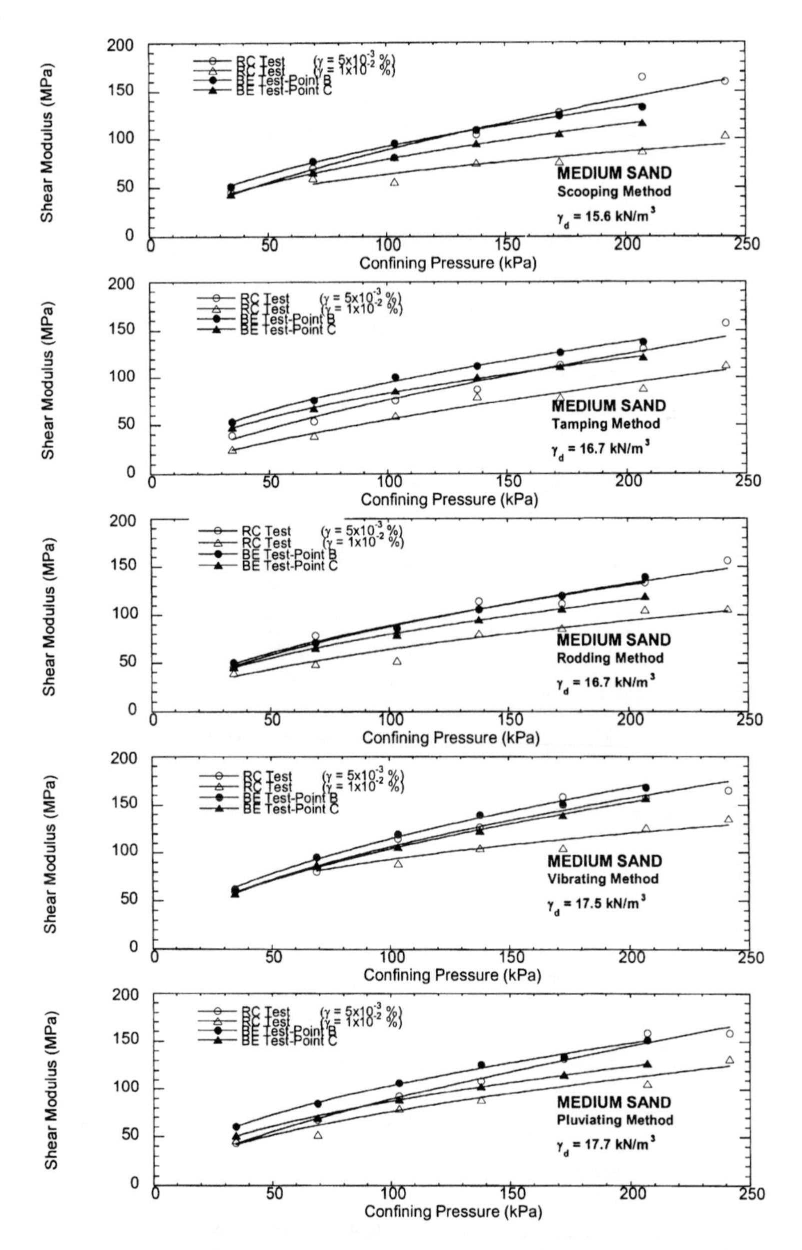

FIG. 3. Shear modulus vs. confining pressure relationship of sands prepared by five methods: scooping, tamping, rodding, vibrating, and pluviating

plate. This P-wave travels along the specimen and is then received by the geophone, which was attached to another aluminum plate located at the bottom of the mold. The travel time of the P-wave from the source to the receiver was recorded by an oscilloscope. Based on the travel time of P-wave and the distance between source and the receiver, the P-wave velocity was calculated. Knowing the density of the specimen and the estimated Poisson's ratio (ν), the Young's modulus and the corresponding shear modulus of the sand specimen for were determined.

Shear moduli of sands obtained from the bender element and the SSG tests are compared with those obtained from the pulse echo tests in Fig. 4. Note that the shear

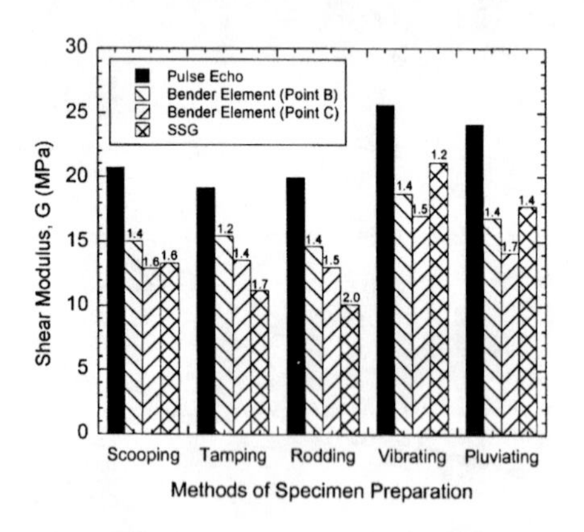

FIG. 4. Comparison of shear modulus of sands prepared by five methods

moduli obtained from the bender elements are extrapolated estimates at the stress level comparable to that induced by the SSG and the pulse echo tests (i.e., 2.6 kPa). The shear moduli obtained from the pulse echo test are consistently higher than those obtained from the bender element tests. However, inconsistency was observed between the shear moduli obtained from the bender elements and the SSG. The ratios of modulus obtained from the pulse echo test to that from the bender elements range 1.2-1.4 for the travel time of S-wave taken at point B and 1.4-1.7 taken at point C, whereas the modulus ratios between the pulse echo and the SSG range 1.2-2.0 (noted above the bars in Fig. 4).

STRAIN LEVELS IN PAVEMENT BASES AND SUBGRADES

The strains induced in the pavement bases and subgrades that are subjected to the typical traffic loadings were compiled from other studies. In general, they can be classified into three main groups: (1) finite-element analysis, (2) large-scale model experiment, and (3) in-situ test section. Table 1 presents the vertical strains in base, subbase, and subgrade layers summarized from various studies. The type of measurement and/or analysis, pavement structure, and loading characteristic employed in each study are also reported herein. Typically, the vertical strains in base, subbase, and subgrade layers are approximately 0.01-0.3%, 0.01-0.7%, and

TABLE 1. Vertical strains in base, subbase, and subgrade layers

References (1)	Method of Estimating Strain (2)	Type of Measurement and/or Analysis (3)	Pavement Structure (4)	Loading Characteristic (5)	Vertical Strain (%)		
					Base (6)	Subbase (7)	Subgrade (8)
Brown & Pappin (1981)	2-D finite-element analysis	Nonlinear	- 50-mm surface - 170-mm base - Subgrade	- 8-kN wheel load - 530-kPa contact pressure	0.15	NA	0.12-0.18
Chen et al. (1986)	3-D finite-element analysis	Linear elastic	- 38-mm, 50-mm, 76-mm & 100-mm surface - 200-mm base - 4293-mm subgrade	- 20-kN & 24-kN wheel loads - 518-kPa, 621-kPa & 759-kPa contact pressures	NA	NA	0.035-0.074
Hardy & Cebon (1993)	Test section	LVDTs	- 150-mm surface - 300-mm base - 914-mm subgrade	- Four-axle vehicle: 29.1-kN steering axle, 40.4-kN-drive axial, 37.9-kN trailer's tandem axle group - Speed: 50 & 80 km/h	NA	NA	0.04-0.07
Marsh & Jewell (1994)	Test section	Vertical strain transducers	- 30-mm surface - 75-mm base - 250-mm subbase - 400-mm subgrade	- 553-kPa & 725-kPa contact pressures	NA	NA	7×10^{-3}-0.012
Chen et al. (1995)	- 2-D &3-D finite-element analysis - Multilayered elastic-based program	Linear & nonlinear	- 76-mm, 152-mm & 229-mm surface - 305-mm base - Subgrade	- 40.5 kN wheel load - 689 kPa contact pressure	NA	NA	0.013-0.13
Pidwerbesky (1995)	Large-scale model experiment	Strain coil sensors	- 25-mm, 35-mm & 85-mm surface - 135-mm, 200-mm & 300-mm base - 200-mm subgrade	- 21-kN, 31-kN, 40-kN & 46-kN wheel load - 550-kPa, 700-kPa & 825-kPa contact pressures	0.09-0.32	NA	0.09-0.35
Dai & Van Deusen (1998)	Test section	LVDTs	- 127-mm & 200-mm surface - 305-mm, 460-mm & 710-mm base - 2690-mm, 2940-mm & 3168-mm subgrade	- Five-axle tractor-trailer: 53.4-kN steering axle, 75.2-kN front axle, 73.9-kN back axle, 69.4-kN front axle & 81.9-kN back axle of tractor tandem - Speed: 16-78 km/h	NA	NA	1.9×10^{-3}-0.019

Helwany et al. (1998)	2-D & 3-D finite-element analysis	Linear & nonlinear	- 150-mm surface - 250-mm base - 650-mm subbase - Subgrade	- 90-kN axle load (45-kN wheel load) - 550-kPa contact pressure	NA	0.012-0.058	NA
Saleh et al. (2003)	Large-scale model experiment	Strain coil sensors	- 25-mm surface - 275-mm base - 1200-mm subgrade	- 40-kN, 50-kN & 60-kN wheel loads - 650-kPa, 700-kPa, 750-kPa, 800-kPa & 850-kPa contact pressures - Speed: 6 km/h	0.02-0.13	NA	0.09-0.35
Tanyu et al. (2003)	Multilayered elastic-based program	- Linear & nonlinear	- 125-mm surface - 255-mm base - 220-mm to 900-mm subbase - 450-mm subgrade	- 35-kN wheel load - 700-kPa contact pressure	NA	0.036-0.77	NA
Tutumluer et al. (2003)	- Large-scale model experiment - 2-D finite-element analysis	- Strain coil sensors - Linear & nonlinear - Isotropic & anisotropic	- 89-mm surface - 203-mm base - 1270-mm subgrade	- 28.9-kN wheel load - 689-kPa contact pressure	0.011-0.062	NA	0.175-0.25
de Pont et al. (2004)	Large-scale model experiment	Strain coil sensors	- 85-mm surface - 200-mm base - Subgrade	- 98-kN axle load (49-kN wheel load) - Speed: 45 km/h	0.06-0.21	NA	0.15-0.31
Huang (2004)	Multilayered elastic-based program	Linear elastic	- 25-mm to 203-mm surface - 102-mm to 406-mm base - Subgrade	- 80-kN axle load (40-kN wheel load) - 690-kPa contact pressure	NA	NA	0.025-0.4

0.002-0.4%, respectively. By assuming that the vertical strains in each layer are in the principal plane, the maximum shear strains in the principal plane are computed by multiplying vertical strains by (1+ν) (Kim and Stokoe 1992). Poisson's ratios (ν) of 0.35, 0.35, and 0.45 were respectively assumed for the base, subbase, and subgrade. The range of maximum shear strains in principle plane are therefore computed to be 0.014-0.41% for base, 0.014-0.95% for subbase, and 0.003-0.58% for subgrade.

RELATIONSHIP BETWEEN SMALL-STRAIN AND LARGE-STRAIN MODULUS

For pavement bases and subgrades, the stress-strain behavior of soil is highly nonlinear and soil modulus may decay with strain by orders of magnitude. A relationship between the small-strain (linear-elastic) modulus (strains less than 10^{-2} %) and nonlinear behavior exhibited by soils at large strains (above 10^{-2} %) must be established. The shear modulus of soil at various shear strain levels for different pavement layers and modulus tests is shown in Fig. 5. Generally, strains in base and

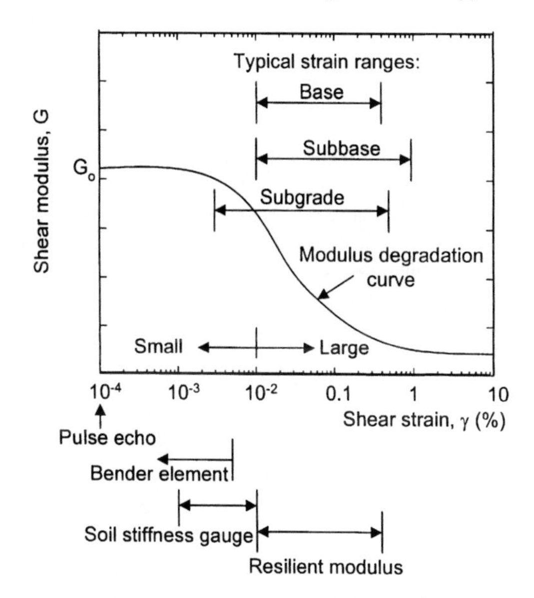

FIG. 5. Typical variation in shear modulus with various shear strain levels

subbase vary from 0.01 to 1%, whereas those in subgrade may vary from 0.003-0.6%. Therefore, the pavement base, subbase, and subgrade layers involve strains at higher levels, i.e., typical strain range of 10^{-2} to 1%, and soil exhibits nonlinear properties. Fig. 5 also shows that the resilient modulus (M_r) test operates within these strain range. However, the measured soil modulus from the bender element and the SSG must be adjusted or corrected to the modulus corresponding to these strain levels.

Since the soil moduli obtained from a variety of tests is likely to be stress- and stain-dependent, the moduli obtained from one test to the others must be effectively compared at the same stress level. A normalized modulus reduction curve (either E/E_o versus log ε_a or G/G_o versus log γ) may be used to assess the variation in moduli with strain amplitude without the effect of state of stress (Seed and Idriss 1970; Hardin and Drnevich 1972; Kokusho et al. 1982; Jardine et al 1986; Sun et al. 1988; Vucetic and Dobry 1991; Tatsuoka and Shibuya 1991; Puzrin and Burland 1996). Once the normalized modulus reduction curve is obtained by using either a database of resonant column test results or a typical modulus degradation scheme (Kim et al. 1997), the elastic modulus of a given soil obtained from either the laboratory or in situ small-strain stiffness measurements can be adjusted to the modulus corresponding to larger strains where the soil exhibits nonlinear behavior. In other words, the strain-dependent modulus of the soil can be predicted using a normalized modulus reduction curve. The modulus at a desired strain level can be determined by combining the measured modulus with the normalized modulus reduction curve as follows:

$$G_r = RF_{\text{data base or deg radation scheme}} \times G_{\text{SSG or BE}} \qquad (2)$$

where G_r is design resilient shear modulus, RF is reduction factor that accounts for shear strain amplitude difference between resilient shear modulus and SSG or BE shear modulus which can be obtained from the database or degradation scheme, and $G_{\text{SSG or BE}}$ is the measured SSG or BE shear modulus. RF is computed as the ratio between the shear modulus corresponding to the shear strain amplitude of resilient modulus test and the shear modulus corresponding to the shear strain amplitude of the SSG or BE. It can be estimated approximately using the modulus degradation curve obtained either from a database of various modulus tests e.g., resonant column, resilient modulus, SSG, BE tests or from the typical modulus degradation scheme proposed by different investigators for soils as mentioned above. The shear strain levels induced by the SSG or BE, which is used to determine the reduction factor can be estimated approximately from Fig. 5. Resilient shear modulus can be converted to axial resilient modulus using $M_r = 2G_r(1+\nu)$. A comparison of moduli obtained from alternative methods with resilient modulus of the same material is given in Sawangsuriya et al. (2003; 2004).

SUMMARY

This paper presents the application of alternate methods which have not been widely used for obtaining moduli for pavement bases and subgrades. Mechanistic design methods require knowledge of the moduli for all pavement materials to determine the pavement design life. Resilient modulus is used in the mechanistic design; however, it can also be estimated from alternative tests such as the bender elements test in the laboratory and the SSG test in the field. A suggested method for adjusting the small-strain modulus to obtain the large-strain resilient modulus is also described in the paper. Although a medium sand was used in this investigation, the general concepts and procedures are applicable to other base and subgrade materials.

REFERENCES

Arulnathan, R., Boulanger, R. W., and Riemer, M. F. (1998). "Analysis of bender element tests." *Geotech. Testing J.*, 21(2), 120-131.

Brown, S. F., and Pappin, J. W. (1981). "Analysis of pavements with granular bases." *Transp. Res. Rec.*, 810, 17-23.

Burland, J. B. (1989). "The ninth Lauritis Bjerrum memorial lecture: 'small is beautiful'-the stiffness of soils at small strains." *Canadian Geotech. J.*, 26, 499-516.

Chen, D.-H., Zaman, M., Laguros, J., and Soltani, A. (1995). "Assessment of computer programs for analysis of flexible pavement structure." *Transp. Res. Rec.*, 1482, 123-133.

Chen, H. H., Marshek, K. M., and Saraf, C. L. (1986). "Effects of truck tire contact pressure distribution on the design of flexible pavements: a three-dimensional finite element approach." *Transp. Res. Rec.*, 1095, 72-78.

Dai, S., and Van Deusen, D. (1998). "Field study of in situ subgrade soil response under flexible pavement." *Transp. Res. Rec.*, 1639, 23-35.

de Pont, J., Thakur, K., Pidwerbesky, B., and Steven, B. (2004). "Validating a whole life pavement performance model." *Transit New Zealand, CAPTIF Research Facility*, 15

Dyvik, R., and Madshus, C. (1985). "Lab measurements of G_{max} using bender elements." *Proc. Advances in the Art of Testing Soil Under Cyclic Conditions*, ASCE, Detroit, MI, 186-196.

Edil, T. B., and Sawangsuriya, A. (2005). "Earthwork quality control using soil stiffness." *Proc. 16th Int. Conference on Soil Mech. and Geotech. Engrg.*, Osaka, Japan (accepted for publication).

Fam, M., and Santamarina, J. C. (1995). "Study of geoprocesses with complementary wave measurements in an oedometer." *Geotech. Testing J.*, 18(3), 307-314.

Fioravante, V., and Capoferri, R. (2001). "On the use of multi-directional piezoelectric transducers in triaxial testings." *Geotech. Testing J.*, 24(3), 243-255.

Hardin, B. O., and Black, W. L. (1968). "Vibration modulus of normally consolidated clay." *J. Soil Mech. and Found. Div.*, ASCE, 94(SM2), 353-369.

Hardin, B. O., and Drnevich, V. P. (1972). "Shear modulus and damping in soils: design equations and curves." *J. Soil Mech. and Found. Div.*, ASCE, 98(SM7), 667-692.

Hardy, M. S. A., and Cebon, D. (1993). "Response of continuous pavements to moving dynamic loads." *J. Engrg. Mech.*, ASCE, 119(9), 1762-1780.

Helwany, S., Dyer, J., and Leidy, J. (1998). "Finite-element analyses of flexible pavements." *J. Transp. Engrg.*, ASCE, 124(5), 491-499.

Huang, Y. H. (2004). Pavement analysis and design. Pearson Prentice Hall, Upper Saddle River, NJ.

Humboldt Mfg. Co. (1999). *Humboldt soil stiffness gauge (geogauge) user guide: version 3.3*, Norridge, IL.

Jardine, R. J., Potts, D. M., Fourie, A. B., and Burland, J. B. (1986). "Studies of the influence of non-linear stress-strain characteristics in soil-structure interaction." *Geotechnique*, 36(3), 377-396.

Jardine, R. J., and Potts, D. M. (1988). "Hutton tension platform foundation: an approach to the prediction of pile behavior." *Geotechnique*, 38(2), 231-252.

Kim, D.-S., and Stokoe, K. H., II (1992). "Characterization of resilient modulus of compacted subgrade soils using resonant column and torsional shear tests," *Transp. Res. Rec.*, 1369, 83-91.

Kim, D.-S., Kweon, G.-C., and Lee, K.-H. (1997). "Alternative method of determining resilient modulus of compacted subgrade soils using free-free resonant column test." *Transp. Res. Rec.*, 1577, 62-69.

Kokusho, T., Yoshida, Y., and Esashi, Y. (1982). "Dynamic properties of soft clays for wide strain range." *Soils and Found.*, 22(4), 1-18.

Mancuso, C., Vassallo, R., and d'Onofrio, A. (2002). "Small strain behavior of a silty sand in controlled-suction resonant column-torsional shear tests." *Canadian Geotech. J.*, 39(1), 22-31.

Mancuso, C., and Vinale, F. (1988). "Propagazione delle onde Sismiche: Teoria e Misura In Sito." *Atti del Convegno del Gruppo Nazionale di Coordinamento per gli Studi di Ingegneria Geotecnica*, Monselice, Rome, 115-138.

Marsh, J. G., and Jewell, R. J. (1994). "Vertical pavement strain as means of weighing vehicles." *J. Transp. Engrg.*, ASCE, 120(4), 617-632.

Pennington, D. S., Nash, D. F. T., and Lings, M. L. (2001). "Horizontally mounted bender elements for measuring anisotropic shear moduli in triaxial clay specimens." *Geotech. Testing J.*, 24(2), 133-144.

Pidwerbesky, B. D. (1995). "Strain response and performance of subgrades and flexible pavements under various loading conditions." *Transp. Res. Rec.*, 1482, 87-93.

Puzrin, A. M., and Burland, J. B. (1996). "A logarithmic stress-strain function for rocks and soils." *Geotechnique*, 46(1), 157-164.

Saleh, M. F., Steven, B., and Alabaster, D. (2003). "Three-dimensional nonlinear finite element model for simulating pavement response: study at Canterbury accelerated pavement testing indoor facility, New Zealand." *Transp. Res. Rec.*, 1823, 153-162.

Sawangsuriya, A., Bosscher, P. J., and Edil, T. B. (2004). "Application of the soil stiffness gauge in assessing small-strain stiffness of sand with different fabrics and densities." *Submitted to Geotech. Testing J.*, ASTM. (contact authors)

Sawangsuriya, A., Edil, T. B., and Bosscher, P. J. (2003). "Relationship between soil stiffness gauge modulus and other test moduli for granular soils." *Transp. Res. Rec.*, 1849, 3-10.

Seed, H. B., and Idriss, I. M. (1970). "Soil moduli and damping factors for dynamic response analyses." *Report EERC 70-10*, Earthquake Engineering Research Center, University of California, Berkeley, CA.

Souto, A., Hartikainen, J., and Özüdoğru, K. (1994). "Measurement of dynamic parameters of road pavement materials by the bender element and resonant column tests." *Geotechnique*, 44(3), 519-526.

Sun, J. I., Golesorkhi, R., and Seed, H. B. (1988). "Dynamic moduli and damping ratios for cohesive soils." *Report EERC-88/15*, Earthquake Engineering Research Center, University of California, Berkeley, CA.

Tanyu, B. F., Kim, W. H., Edil, T. B., and Benson, C. H. (2003). "Comparison of laboratory resilient modulus with back-calculated elastic moduli from large-scale model experiments and FWD tests on granular materials." *Resilient Modulus Testing for Pavement Components, ASTM STP 1437*, West Conshohocken, PA, 191-208.

Tatsuoka, F., and Shibuya, S. (1991). "Deformation characteristics of soils and rocks from field and laboratory tests." *Proc. 9th Asian Regional Conference on Soil Mech. and Found. Engrg.*, Bangkok, Thailand, 2, 101-170.

Tatsuoka, F., Siddiquee, M. S., Park, C. S., Sakamoto, M., and Abe, F. (1993). "Modeling stress-strain relations of sand." *Soils and Found.*, 33(2), 60-81.

Thomann, T. G., and Hryciw, R. D. (1990). "Laboratory measurement of small strain shear modulus under K_o conditions." *Geotech. Testing J.*, 13(2), 97-105.

Tutumluer, E., Little, D. N., and Kim, S.-H. (2003). "Validated model for predicting field performance of aggregate base courses." *Transp. Res. Rec.*, 1837, 41-49.

Vucetic, M., and Dobry, R. (1991). "Effect of soil plasticity on cyclic response." *J. Geotech. Engrg.*, ASCE, 117(1), 89-107.

Zeng, X., and Ni, B. (1998). "Application of bender elements in measuring G_{max} of sand under K_o condition." *Geotech. Testing J.*, 21(3), 251-263.

THE USE OF FLY ASH FOR IN-SITU RECYCLING OF AC PAVEMENTS INTO BASE COURSES

Bruce W. Ramme[1], Haifang Wen[2], Tarun R. Naik [3] and Rudolph N. Kraus[4]

ABSTRACT: Class C fly ash is a coal combustion product normally produced from lignite or sub-bituminous coal obtained as a result of the power generation process. In recent years, efforts were taken to incorporate self-cementing fly ash into full-depth reclaimed (FDR) asphalt pavements to improve the structural capacity of asphalt pavement base layers. In this study, an existing asphalt pavement in County Trunk Highway (CTH) "JK" in Waukesha County, Wisconsin was pulverized in place and mixed with fly ash and water to function as a base course. To evaluate the contribution of fly ash to the structural performance of the pavement, nondestructive deflection tests were performed using a KUAB 2M falling weight deflectometer (FWD) on the outer wheel path four days, one year, and two years after construction. The modulus of the fly ash stabilized FDR base course increased by 49% one year after construction, and by 83% two years after construction. The structural capacity of the fly ash stabilized FDR base course in CTH "JK" also increased significantly as it ages, due to the pozzolanic and cementitious reactions. The results of this study indicate that the FDR mixtures with self-cementing fly ash can provide an economical method of recycling flexible pavements and reduce the need for expensive new granular base courses for road reconstruction.

INRODUCTION

The Full-depth Reclamation (FDR) of asphalt pavements process consists of in situ pulverization of the existing asphalt layer along with the existing aggregate base and sometimes subgrade soils to form a base for a new asphalt overlay. FDR is becoming a widely used rehabilitation technique by highway agencies. There are two reasons: (1) compared to conventional methods, such as milling and patching, FDR is a more effective way to eliminate the reflective cracking of asphaltic overlays and

---

[1] Manager of Environmental Land Quality, We Energies, 333 West Everett Street, Milwaukee, WI 53290.

[2] Transportation Engineer, Bloom Consultants, LLC, 10001Innovation Drive, Suite 200, Milwaukee, WI 53226.

[3] Professor and Director of the UWM center for By-Product Utilization, University of Wisconsin-Milwaukee, Milwaukee, WI 53201.

[4] Assistant Director of the UWM center for By-Product Utilization, University of Wisconsin-Milwaukee, Milwaukee, WI 53201.

improve the pavement base condition; and (2) FDR utilizes 100% of the existing materials and is a cost-effective technique, considering the shortage of available quality aggregates and more stringent environmental regulations. FDR projects have been performed successfully since the early 1980's in states such as Kansas, Oregon, California and New Mexico.

The pulverized materials without additives could be used as pavement base course after compaction and grading (Kearney and Huffman, 1999; Wilson et al., 1998). However, additives are often used to stabilize the pulverized base course. The available additives in common practice include water, cement, asphalt emulsion, hydrated lime, and fly ash (Cross and Fagged, 1995; Cross and Young, 1997; Malice et al. 2002).

Class C fly ash, a coal combustion product from lignite or sub-bituminous coal obtained as a result of the power generation process, has been used extensively over a wide range of construction applications. Its self-cementitious and pozzolanic properties are valuable in developing strength of concrete or other mixtures containing fly ash. Each year, approximately 68 million tons of fly ash are produced in the U.S.A (American Coal Ash Association, 2002). About 46 million tons were placed in landfills resulting in significant land purchase costs, landfill costs, and potential environmental issues. It has been reported that some FDR mixes with asphalt emulsion were unstable (Cross and Young, 1997) and cement stabilized base course was prone to cracking (Malice et al., 2002). Therefore, it is environmentally friendly and cost-effective to utilize self cementing Class C fly ash to stabilize FDR mixes that have not exhibited these deficiencies. Cross Fagged (1995) studied the use of Class C fly ash to stabilize Cold In-place Recycled base course and concluded fly ash could be a viable additive. However, Cross and Young (1997) reported that the use of high fly ash application rate could result in the tendency of pavement to crack.

Since Class C fly ash is a pozzolanic material, when compared to Portland cement, mixtures with fly ash have a long-term strength development process. Use of appropriate fly ash contents to stabilize FDR mixes can reduce the brittle behavior of the base course and still provide enough support for the long-term performance of asphalt overlays. Therefore, this study investigated the stabilization of FDR mixes with relatively low fly ash contents in a road with moderate traffic volume. This study is evaluating the strength development and multiple-year field performance of a Class C fly ash stabilized FDR pavement base course in Wisconsin.

PROJECT DESCRIPTION

County Trunk Highway (CTH) JK in Waukesha, Wisconsin was selected as a test section. It was reconstructed in October of 2001, using FDR materials as pavement base course. Fly ash was used to stabilize the FDR pavement base course in place for CTH JK. CTH JK is located in Waukesha County, Wisconsin and the project segment runs between CTH KF and CTH K, with a project length of 1,009 m (3,310 ft.). It is a two-lane road with an average daily traffic (ADT) count of 5,050 vehicles in year 2000 and a projected ADT of 8,080 in design year 2021. The existing pavement structure consisted of approximately a 127 mm (5") asphalt concrete surface layer and a 178 mm (7") granular base course. The new pavement structure consists of a 127 mm (5") asphalt concrete layer and a 305 mm (12") Class C fly ash stabilized FDR base course. The truck percentage on CTH JK in 2000 was 5%.

LABORATORY EXPERIMENT AND CONSTRUCTION

A laboratory mix analysis to evaluate the stabilization potential of pulverized pavement material with Class C fly ash was conducted. A field sample of the existing asphalt pavement and underlying aggregate base was obtained. The results of the grain size analysis on the pulverized materials indicated a sand and gravel mixture with trace fines. The analysis showed that the sample contained 68% gravel size particles (larger than #4 sieve), 26% sand size particle (between #4 and #200 sieves), and 6% silt size (between #200 sieve and a size of 0.005 mm) and clay size (between 0.005 mm and 0.001 mm) particles. Based on published technical literature (Cross and Fagged, 1995; Cross and Young, 1997; Malice et al. 2002), a laboratory evaluation of fly ash stabilized FDR material was performed at two fly ash contents, 6 and 8% by dry weight of FDR materials. These two relatively low application contents were selected to reduce the potential of cracking of the stabilized mixtures. Laboratory analysis of the fly ash stabilized materials was conducted in accordance with ASTM C593. Moisture-Density (ASTM D1557) and Moisture-Strength (ASTM D1633) relationships of specimens compacted in a 101.6mm (4") diameter mold were obtained. Results of the moisture density relationship tests on the pulverized asphalt pavement indicated a maximum dry density of 2.27 g/cm^3 (141 p/c.f.) at an optimum moisture level of 5.0%. In addition, the moisture density relationship tests on the recycled asphalt pavement material with 6 and 8% fly ash indicated maximum dry densities of 2.28 and 2.29 g/cm^3 (142 and 143 p/c.f.) at optimum moisture contents of 5.5%, respectively. Maximum unconfined compressive strengths of 1.72 MPa (250 psi) and 2.62 MPa (380 psi) at optimum moisture contents of 5% were obtained after seven day curing for 6% and 8% application rates, respectively.

The pulverized mixes were compacted and graded to form the base for a 127 mm (5") thick new asphalt overlay. Construction of CTH JK consisted of pulverization, application of 8% fly ash and 5% water, compaction and grading, and placement of the new asphalt overlay. For detailed construction procedures, the readers are referred to (Wen et al., 2003).

FIELD PERFORMANCE EVALUATION

The nondestructive deflection testing is one of the primary techniques for determining the in situ structural capacities of pavement. The 1993 *AASHTO Guide for Design of Pavement Structures* (AASHTO Guide for Design of Pavement Structures, 1993) describes Falling Weight Deflectometer (FWD) testing as a means of evaluating the conditions of existing pavement. The NCHRP Project 1-37A also proposes the use of FWD for existing pavement evaluation (Schwartz and El-Basyouny, 2002). In this study the FWD was used to evaluate the performance of fly ash stabilized FDR.

KUAB 2M Falling Weight Deflectometer (FWD) tests were conducted to evaluate the field performance of CTH JK. The impact load used in this study was approximately 40KN (9000 lbs). The pavement surface deflections were recorded by seven sensors located at 0, 0.3 m (12"), 0.46 m (18"), 0.61 m (24"), 0.91 m (36"),

1.22 m (48"), and 1.52 m (60") from the center of loading plate. FWD tests were conducted on CTH JK four days after construction (2001), one year after construction (2002), and two years after construction (2003). The FWD tests were performed at an interval of 30.5 m (100 ft).

DEFLECTION

The pavement deflections under the impact load indicate the structural capacities of existing pavement, including subgrade, base course and surface layer. Deflections may be either correlated directly to pavement performance or used to determine the in situ material characteristics of the pavement layers.

The average deflections measured by the sensors are shown in Table 1.

TABLE 1. Average Deflection Measurements in Year 2001, 2002 and 2003

Year		D_0, mm	$D_{0.3}$, mm	$D_{0.46}$, mm	$D_{0.61}$, mm	$D_{0.91}$, mm	$D_{1.22}$, mm	$D_{1.52}$, mm
Year 2001	Mean	0.234	0.177	0.152	0.129	0.099	0.074	0.056
	Std. Dev.	0.045	0.037	0.031	0.027	0.023	0.019	0.016
	C.V. (%)	19.3	20.6	20.7	21.1	23.2	25.5	27.9
Year 2002	Mean	0.138	0.115	0.106	0.094	0.075	0.062	0.049
	Std. Dev.	0.029	0.024	0.022	0.020	0.017	0.015	0.013
	C.V. (%)	19.6	19.9	19.5	19.4	20.6	22.3	24.2
Year 2003	Mean	0.151	0.125	0.110	0.098	0.080	0.065	0.051
	Std. Dev.	0.028	0.025	0.022	0.020	0.017	0.015	0.012
	C.V. (%)	18.5	19.8	20.0	20.2	21.7	23.1	24.3

The average deflections measured in year 2002 and 2003 were significantly lower than those in year 2001; the deflections in 2003 were slightly higher than those in 2002. The improvement was partly due to the fact that the pavement surface temperature was 16.7°C (62°F) in 2001, 7.2°C (45°F) in 2002, and 8.3°C (47°F) in 2003 at the time of testing. The study of flexible pavements in Texas by Chen et al. (2000) shows that only the deflections at a radial distance of 0 and 203 mm (8") are significantly affected by temperature. Park et al. (2002), however, concluded that the radial distance affected by temperature, is dependent on the thickness of asphalt concrete layer. The effective radial distance for deflection affected by temperature was calculated using the following relationship (Park et al., 2002):

$$Deff = 4.75\,Hac - 413 \quad (1)$$

Where: $Deff$ = effective radial distance affected by temperature, mm, and
Hac = Asphalt Concrete (AC) layer thickness, mm.

In this study, the thickness of the AC layer is 127 mm (5"). Therefore, the effective radial distance affected by temperature is about 190 mm (7.5"). Only the deflections measured at the center of loading plate, D_0, were affected by the temperature in this study. To meaningfully compare the pavement behavior in 2001, 2002 and 2003, the deflections have to be corrected to the reference temperature of 20°C (68°F) in this study. It is well known that the measured pavement surface temperature has to be corrected to that at mid-depth of the asphalt concrete layer to be representative of the effective temperature. The 1993 *AASHTO Guide for Design of Pavement Structures* presents the temperature correction procedure. However, it has been reported that that temperature correction method is inaccurate and impractical (Inge and Kim, 1995). Inge and Kim (1995) developed the temperature correction method, based on the data collected in North Carolina. Another method was developed by Harichandran et al. (2001) for flexible pavement in Michigan. The method which has been developed based on the national, instead of regional data, is BELLS3 by Lukanen et al. (2000). Lukanen et al. (2000) used the data collected in the Long-term Pavement Performance (LTPP) program. This study examined the above temperature correction methods. The results of temperature correction for CTH JK pavement are shown in Table 2.

TABLE 2. Temperature Correction of CTH JK Pavement

Correction Method by		Kim et al.	Harichandran et al.	Lukanen et al. (BELLS3)
Data Source		North Carolina	Michigan	LTPP
Measured	16.6°C (2001)	19.2°C	15.3°C	13.6°C
	7.2°C (2002)	9.1°C	5.8°C	5.2°C
	8.3°C (2003)	10.3°C	6.9°C	6.3°C

The corrected temperatures at the middle of the asphalt layer using the North Carolina method were higher than the measured pavement surface temperatures; while the other two methods, Michigan method and the BELLS3 method, yielded an opposite trend. It was seen that the effective temperatures from the Michigan method, 15.3°C (59.54°F) in 2001, 5.8°C (42.4°F) in 2002 and 6.9°C (44.4°F) in 2003, were close to those from BELLS3 method, 13.6°C (56.5°F) in 2001, 5.2°C (41.4°F) in 2002 and 6.3°C (43.3°F) in 2003. Considering the climatic similarity between Michigan and Wisconsin, it was decided to use the corrected temperature from the Michigan method.

The next step is to correct the measured deflections at the effective temperature at the middle of the asphalt layer to those at the reference temperature of 20°C (68°F). Several deflection correction methods were developed by Park et al. (2002), Chen et al. (2000), and Lukanen et al. (2000), using data from pavements in North Carolina, in Texas, and LTTP, respectively. The deflection correction coefficients from the above methods vary significantly, as shown in Table 3. Again, considering the climatic difference between Wisconsin and North Carolina or Texas, it was decided to use the method developed by Lukanen et al. (2000).

TABLE 3. Deflection Correction Coefficient of CTH JK Pavement

Correction Method by		Kim et al.	Chen et al.	Lukanen et al.
Data Source		North Carolina	Texas	LTPP
Correction Coefficient	2001	1.08	1.28	1.14
	2002	1.26	2.87	1.42
	2003	1.24	2.48	1.39

The deflection curves in 2001 and 2002, after the temperature correction, are presented in Figure 1. The deflections in 2003 are fairly close to those in 2002, after temperature correction. It was noted that the deflections measured by sensor located at 1.52 m (60") from the center of loading plate, $D_{1.52}$, in 2002 and 2003 were slightly lower than those in year 2001. Knowing that $D_{1.52}$ is the measurement of the deflection of only the subgrade, it is believed that the slightly reduced deflection of subgrade might be ascribed to further compaction of subgrade by traffic. It appears that the corrected deflections in 2002 and 2003 were significantly lower than those in 2001. Therefore, it is inferred that that the fly ash stabilized cold in-place recycled asphalt base course gained strength significantly between the time of testing in 2001 and in 2002 and 2003, due to pozzolanic and cementitious reactions, and thus reduced the pavement deflection under the loading.

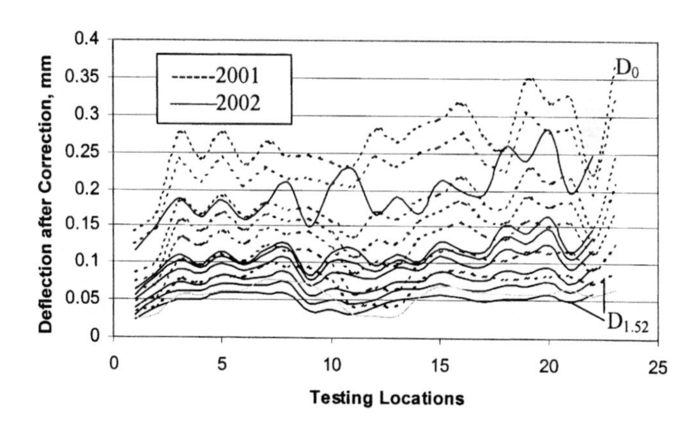

FIG. 1. Pavement Deflection from FWD Test after Temperature Correction

BACKCALCULATION OF LAYER MODULUS

The measured deflection data was used to backcalculate the properties of each pavement layer. Modulus 5.1 and Michback programs were used in the preliminary analysis. It was found that both programs yielded close results. It was decided to continue the backcalculation using only the Michback program. The average moduli of the materials in the asphalt layer, the fly ash stabilized base course and the subgrade were 6.68GPa (968ksi), 1.24GPa (180ksi), and 0.1GPa (14.5ksi), respectively, at the time of testing in 2001; 25.3GPa (3,670ksi), 1.84GPa (267ksi) and 0.1GPa (14.5ksi) in 2002; and 14.3GPa (2,074ksi), 2.27GPa (329ksi), and 0.11GPa (16.0ksi) in 2003, respectively. The modulus of fly ash stabilized FDR base course increased from 1.24GPa (180ksi) in 2001, to 1.84GPa (267ksi) (by 49%) in 2002, and to 2.27GPa (329ksi) (by 83%) in 2003. The results indicated that the structural capacity of the fly ash stabilized CIR recycled asphalt base course developed significantly within two years after construction. This is due to the pozzolanic and cementitious reactions in the mixes containing Class C fly ash, as stated above.

STRUCTURAL NUMBER

The structural number of the pavement was backcalculated from surface deflection, as follows (Crovetti, 1998):

$$SN = \left[1.49 \times (ET)^3\right]^{1/3} \quad (2)$$

$$Log_{10}(ET)^3 = 5.03 - 1.309 Log_{10}(AUPP) \quad (3)$$

$$AUPP = \frac{1}{2}(5D_0 - 2D_{0.3} - 2D_{0.61} - D_{0.91}) \quad (4)$$

where: SN = structural number of pavement, mm,
ET^3 = flexural rigidity of pavement, mm,
AUPP = area under the pavement profile, mm, and
D_i = surface deflection, mm.

Figure 2 shows the structural number SN at individual test locations in year 2001, 2002, and 2003, obtained from the corrected deflections. The increased SN indicates that the structural capacity of CTH JK pavement was improved significantly. The structural coefficient for fly ash stabilized CIR material, α_2, was calculated as follows:

$$\alpha_2 = \frac{SN - \alpha_1 h_{HMA}}{h_{base}} \quad (5)$$

Where: h_{HMA} = thickness of HMA layer, mm, and
h_{base} = thickness of base course, mm.

There are two approaches to backcalculate the structural coefficient of Class C fly ash stabilized FDR base course. The first approach uses the structural number from corrected deflections and the structural coefficients from the corrected asphalt concrete layer modulus at reference temperature, 20°C. The other is based on the structural number from measured deflections and structural coefficients from backcalculated asphalt concrete layer modulus without correction.

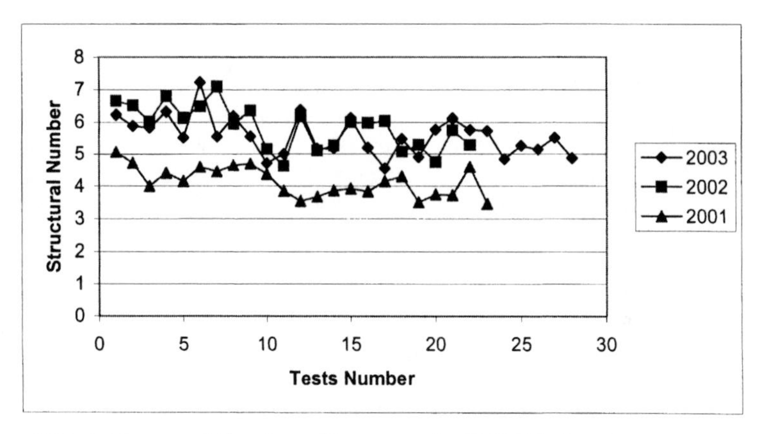

FIG. 2. Structural Number of Pavement in CTH JK after Temperature Correction

In the first approach, backcalculated modulus of the asphalt layer at test temperature was reduced to that at reference temperature. Four correction methods of asphalt layer modulus were examined in this study. They are developed by Kim et al. (1995), Chen et al. (2000), Harichandran et al. (2001), and Lukanen et al. (2000). The results for modulus temperature correction were shown in Table 4. It is seen that the corrected moduli vary significantly. The results from Harichandran et al. and Lukanen et al. are significantly larger than those from Kim et al. and Chen et al., especially at low temperature. The corrected moduli in 2002 and 2003 were not comparable to those in 2001. The reason for the significant discrepancy might be: (1) inaccuracy of modulus correction methods; (2) aging of asphalt concrete materials; and (3) further densification of asphalt concrete layer under traffic. Without knowing the exact reason, it was decided to use the second approach to backcalculate the structural coefficient of the base course containing fly ash.

The structural coefficient of asphalt concrete layer was calculated based on the 1993 *AASHTO Guide for Design of Pavement Structures*, as follows:

$$\alpha_1 = 0.40 \times \log\left(\frac{E}{3000Mpa}\right) + 0.44 \tag{6}$$

where E is the laboratory resilient modulus of asphalt concrete, Pascal.

TABLE 4. Temperature Correction of Asphalt Concrete Modulus

Correction Method by		Kim et al.	Harichandran et al.	Chen et al.	Lukanen et al. (BELLS3)
Data Source		North Carolina	Michigan	Texas	LTPP
Measured	6.68GPa (2001)	4.9GPa	5.3GPa	4.8GPa	5.4GPa
	25.3GPa (2002)	10.3GPa	12.3GPa	8.0GPa	13.4GPa
	14.3GPa (2003)	6.23GPa	7.9GPa	3.3GPa	7.4GPa

According to 1993 *AASHTO Guide for Design of Pavement Structures*, the backcalculated layer modulus of asphalt concrete could be up to three times higher than the resilient modulus obtained in the laboratory. Therefore, the backcalculated modulus value of the asphalt layer was converted into resilient modulus and was input in Equation 6. The structural coefficient obtained form Equation 6 was input in Equation 5.

A structural coefficient of 0.16 was obtained for the fly ash stabilized base course in CTH JK at the time of testing in 2001, 0.23 in 2002, and 0.245 in 2003. Since layer coefficient is a measure of the relative ability of the material to function as a structural component of the pavement, the increase of layer coefficient indicates the improvement of structural capacity of the fly ash stabilized FDR base course in CTH JK. From the standpoint of structural number only, the increase of structure number will result in an allowable traffic increase of 130%.

DISTRESS SURVEY

Condition surveys were conducted to assess the physical condition and distress buildup of the pavement in CTH JK one year and two years after the construction. Of particular interest in these surveys was to identify the possible reflective cracking that was due to the contraction of base course that propagates through the asphalt layer. As seen in Figures 3 and 4, no cracking was observed in the pavement of CTH JK.

The rutting of pavement was surveyed in the inner wheel path of pavement in CTH JK. The rut depth was measured using a straightedge and a gage. It was found that the no rutting happened in pavement of CTH JK. To better visualize the rutting, water was placed on the pavement surface of CTH JK. No accumulation of water was observed within the wheel path, indicating that the materials under the asphalt concrete layer provide a strong support to the load of traffic.

FIG. 3. CTH JK Pavement in 2002

FIG. 4. CTH JK Pavement in 2003

CONCLUSIONS

FDR is becoming a widely used rehabilitation technique by highway agencies. It is also environmentally friendly and cost-effective to utilize Class C fly ash to stabilize FDR pavements. CTH JK in Waukesha, Wisconsin was selected as a test section. A laboratory mix analysis to evaluate the stabilization potential of recycled pavement material with Class C fly ash was conducted. Pavement performance of CTH JK was evaluated using the FWD test after construction, one year, and two years after construction. The modulus of fly ash stabilized FDR base course increased from 1.24GPa (180ksi) in 2001, to 1.84GPa (267ksi) (by 49%) in 2002, and to 2.27GPa (329ksi) (by 83%) in 2003. A structural coefficient of 0.16 was obtained for the fly ash stabilized base course in CTH JK at the time of testing in 2001, 0.23 in 2002, and 0.245 in 2003. The increase of layer coefficient indicates the improvement of structural capacity of the fly ash stabilized FDR base course in CTH JK. It is believed this is due to the long-term pozzolanic and cementitious reactions in the mixtures containing Class C fly ash. No distresses occurred in CTH JK two years after construction. The results of this study at this stage indicate that the CIR stabilization with self-cementing fly ash provide a sustainable economical method of recycling flexible pavements and reduce the need for expensive new granular base courses for road reconstruction.

ACKNOWLEDGEMENT

We would like to thank Mr. Tom Kiernan of Lafarge North America for the support we received. Our special thanks go to Mr. Steve Krebs, Mr. Michael Malaney, Ms. Colleen Coolly, and Mr. Craig Villas of Wisconsin Department of Transportation who helped us by providing the FWD testing for this project.

REFERENCES

Kearney, E. J. and Huffman, J., "Full-depth Reclamation Process." Transportation Research Record, No. 1684, Transportation Research Board, National Research Council, Washington D.C., 1999, pp203-209.

Wilson, J., Fischer, D., and Martens, K., "Pulverize, Mill & Relay Asphaltic Pavement & Base Course." Construction Report, No. WI-05-98, Wisconsin Department of Transportation, 1998.

Cross, S.A. and Fagged, GA, "Fly Ash in Cold Recycled Bituminous Pavement." Transportation Research Record, No 1486, Transportation Research Board, National Research Council, Washington D.C., 1995, pp.49-56.

Cross, S.A. and Young, D., "Evaluation of Type C Fly Ash in Cold In-Place Recycling." Transportation Research Record, No 1583, Transportation Research Board, National Research Council, Washington D.C., 1997, pp.83-90.

Malice, R.B., Boner, D.S., Bradbury, R.L., Andrews, J.O., Kandhal, P.S., and Kearney, E.J., "Evaluation of Performance of Full-Depth Reclamation Mixes." Transportation Research Record, No 1809, Transportation Research Board, National Research Council, Washington D.C., 2002, pp.199-208.

American Coal Ash Association, "2001 Coal Combustion Product (CCP) Production and Use." 2002.

Wen, H., Tharaniyil, M., and Ramme, B., "Investigation of Performance of Asphalt Pavement with Fly Ash Stabilized Cold In-Place Recycled Base Course." Transportation Research Record, No 1819, Vol. 2, Transportation Research Board, National Research Council, Washington D.C., 2003, pp.27-31.

AASHTO Guide for Design of Pavement Structures. American Association of State Highway and Transportation Officials, Washington, D.C., 1993.

Schwartz, C. and El-Basyouny, M., "Flexible Pavement Design and Rehabilitation Procedure." Presented at 2002 Guide for Mechanistic Pavement Design Workshop, Transportation Research Board Annual Meetings, Washington D.C., January, 2003

Chen, D.H., Bilyeu, J., Lin, H.H., and Murphy, M., "Temperature Correction on Falling Weight Deflectometer Measurements." Transportation Research Record, No 1716, Transportation Research Board, National Research Council, Washington D.C., 2000, pp.30-39.

Park, H.M., Kim, Y.R., and Park, S., "Temperature Correction of Multi-Load Level FWD Deflection." Presented at 81 TRB Annual Meeting, Transportation Research Board, National Research Council, Washington D.C., 2002.

Inge, E.H. and Kim, Y.R., "Prediction of Effective Asphalt Layer Temperature." Transportation Research Record, No 1473, Transportation Research Board, National Research Council, Washington D.C., 1995, pp.93-100.

Harichandran, R.S., Buch, N., and Baladi, G.Y., "Flexible Pavement Design in Michigan." Transportation Research Record, No 1778, Transportation Research Board, National Research Council, Washington D.C., 2001, pp.100-106.

Lukanen, E. O., Stubstad, R. N., and Briggs, R. Temperature Predictions and Adjustment Factors for Asphalt Pavement. Report FHWA-RD-98-085. FHWA, U.S. Department of Transportation, 2000.

Crovetti, J., "Design, Construction, and Performance of Fly Ash Stabilized CIR Asphalt Pavements in Wisconsin", Wisconsin Electric-Wisconsin Gas, Milwaukee, Wisconsin, 1998.

Kim, Y.R., Bibbs, B.O., and Lee, Y.C., "Temperature Correction of Deflections and Backcalculated Asphalt Concrete Moduli." Transportation Research Record, No 1473, Transportation Research Board, National Research Council, Washington D.C., 1995, pp.55-62.

USE OF ELASTIC AND ELECTROMAGNETIC WAVES TO EVALUATE THE WATER CONTENT AND MASS DENSITY OF SOILS: POTENTIAL AND LIMITATIONS

Bashar Alramahi[1], Khalid A. Alshibli[2] and Dante Fratta[3]

Abstract

The use of elastic and electromagnetic waves to monitor soil properties provides complementary information about stiffness and phase distribution. This paper presents a methodology for monitoring electromagnetic (ELM) and elastic wave parameters as the soil is being inundated with water. The approach helps relating volumetric water content to stiffness and hints to the use of the technique for non-destructive evaluation of in situ water content and mass density of soils. The methodology is tested using a simple numerical study that incorporates errors in the simulated measurement of volumetric water content and P-wave velocity. Results show that even with the presence of errors, the inverted results match the simulated values of mass density and water content. The development of such a technique opens an opportunity for the advancement in measurement practices over time-consuming and destructive techniques (e.g., sand cone and water balloon methods) and for the replacement of regulated materials (i.e., nuclear moisture-density apparatus).

1. INTRODUCTION

Field density and water content measurements are essential in the design and evaluation of many civil engineering projects. These parameters are used as criteria for the acceptance of earth works projects including embankments, earth data, road sub-bases and clay liners in land fields. Furthermore, geotechnical engineers base their design on the properties obtained during the proper compaction of soils. Research in geotechnical engineering has shown that the soil structure obtained by compacting the soil wet or dry of optimum yields very different behavior response (see for example Daniel and Benson 1990).

[1] Graduate Student. Civil and Environmental Engineering. Louisiana State University. Baton Rouge, LA 70803, Email: balram1@lsu.edu

[2] Associate Member. Joint Assistant Professor, PE. Civil and Environmental Engineering. Louisiana State University and Southern University, Baton Rouge, LA 70803, Email: alshibli@lsu.edu

[3] Assistant Professor, PE. Civil and Environmental Engineering and Geological Engineering Program. University of Wisconsin-Madison. Madison, WI 53706, Email: fratta@wisc.edu

Traditional and proven field practices for the evaluation of the compacted soil density and water content include the use of simple methods, such as the sand cone and water balloon methods with the removal of soil for the evaluation of water content (ASTM D1556-00; ASTM D2167-94). The methods are widely accepted but are time consuming and require the collection of specimens that makes the process destructive.

A very well-known alternative is the use of the nuclear density gauge (ASTM D3017-04) to determine both the in situ field density and water content. This instrument allows for the rapid determination of these parameters by monitoring the attenuation of the energy generated by a radiation source. The radioactive nature of the instrument limits its transport and requires certification of the personnel to be able to use it.

Due to limitations in the time to collect large number of specimens or in the safety-related issues of the instruments, both university researchers and research & development companies are trying to design and test new methodologies for the rapid evaluation of in situ water content and mass density. There are different efforts in the development of alternative techniques and instruments that may accomplish this task with different levels of success. For example, the GeoGauge (ASTM D6758-02) measures the small-strain stiffness of soils at depths one to two times the depth of its base. This instrument can be used as an indirect measure of the quality of compacted materials (Abu-Farsakh et al. 2004; Sawangsuriya et al. 2003; 2004). Another instrument is the Soil Quality Indicator (SQI) that is being developed to evaluate the quality of compacted soils (including mass density). The SQI instrument uses electrical impedance spectroscopy and artificial neural network algorithms to back-calculate the soil density and moisture content (TransTech 2005).

The most promising technique is the Purdue TDR Method developed by Drnevich and co-workers (Siddiqui and Drnevich 1995; Lin et al. 2000; ASTM D6780-02). The Purdue TDR method utilizes data collected with the Time Domain Reflectometry (TDR) technique to estimate the soil water content and density. It involves driving four spikes into the soil surface using a template (Drnevich *et al.* 2003; Yu and Drnevich 2004). Then, a multiple-rod-head-probe TDR system is placed on the top of the spikes to measure the electromagnetic wave properties. The measurement procedure also includes extracting a soil specimen, placing it in a compaction mold, and measuring electromagnetic wave properties as a way to calibrate the measurements. Based on the two sets of measurements, the water content and the density are calculated.

However, it is not clear why some methods work while some others do not. That is, is the information collected by one technique meaningful enough to evaluate all the parameters? And if not, how do the different non-destructive techniques solve this problem? This paper attempt to answer these questions and presents evidence that an alternative procedure may be used to evaluate the mass density and water content by combining dielectric permittivity and P-wave velocity of soils as the water content is increased.

2. BASIC DEFINTIONS

The soil density ρ is defined as (Figure 1)

$$\rho = (1-n)\cdot\rho_s + n\cdot S_r\cdot\rho_w + n\cdot(1-S_r)\cdot\rho_a = (1-n)\cdot\rho_s + \theta_v\cdot(\rho_w-\rho_a) + n\cdot\rho_a \quad (1)$$

where ρ_s, ρ_w and ρ_a are the density of the solid, water, and air phases; and n, S_r and θ_v are the porosity, degree of saturation and volumetric water content. Knowing that the density of air is much smaller than the density of soil and water phases, Equation 1 may be simplified to:

$$\rho = (1-n)\cdot\rho_s + n\cdot S_r\cdot\rho_w = (1-n)\cdot\rho_s + \theta_v\cdot\rho_w \quad (2)$$

Equation 2 shows that to be able to evaluate the mass density of soils, the engineer should be able to independently measure n, S_r (or θ_v) and ρ_s (or G_s). G_s ($=\rho_s/\rho_w$) is the specific gravity of solids and varies in a narrow range of values (from 2.6 to 2.8 for most soil minerals). The mass density of water is assumed to be known.

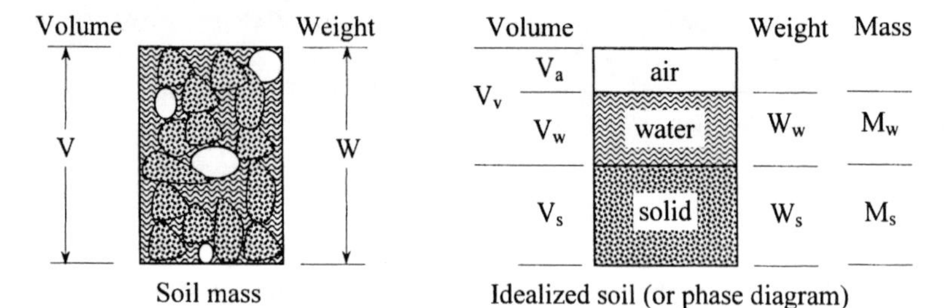

Soil mass — Idealized soil (or phase diagram)

Figure 1. Phase diagram: Definitions

3. ALTERNATIVE METHODOLOGIES: POTENTIALS AND LIMITATIONS

Electric conductivity σ and relative dielectric permittivity κ are expressed as functions of the porosity n, degree of saturation S_r (or porosity and volumetric water content θ_v), and the inherent properties of the different phases in the soil:

$$\sigma^\beta = (1-n)\cdot\sigma_s^\beta + n\cdot S_r\cdot\sigma_w^\beta + n\cdot(1-S_r)\cdot\sigma_a^\beta = (1-n)\cdot\sigma_s^\beta + \theta_v\cdot(\sigma_w^\beta-\sigma_a^\beta) + n\cdot\sigma_a^\beta \quad (3)$$

$$\kappa^\beta = (1-n)\cdot\kappa_s^\beta + n\cdot S_r\cdot\kappa_w^\beta + n\cdot(1-S_r)\cdot\kappa_a^\beta = (1-n)\cdot\kappa_s^\beta + \theta_v\cdot(\kappa_w^\beta-\kappa_a^\beta) + n\cdot\kappa_a^\beta \quad (4)$$

where σ_s, σ_w, and σ_a are the electrical conductivity of solid, water and air phases respectively, κ_s, κ_w, and κ_a are the relative dielectric permittivity of solid, water and air phases respectively, and β is a experimentally determined parameter. Equations 3 and 4 do not include surface conduction, tortuosity and double layer polarization that are important in some soils (in particular in fine soils – see for example Santamarina et al. 2001).

The rapid analysis of Equations 1, 3 and 4 shows how similar these equations are. Then, it would be natural for many to try to use electromagnetic property-based geophysical and non-destructive measurements to invert for mass density and water content. Furthermore, Equations 3 and 4 may be simplified by using typical values. For example, using σ_s=0 S/m, σ_a=0 S/m, and κ_a=1:

$$\sigma^{\beta} = n \cdot S_r \cdot \sigma_w^{\beta} = \theta_v \cdot \sigma_w^{\beta} \quad \text{or} \quad \sigma = \sqrt[\beta]{n \cdot S_r} \cdot \sigma_w = \sqrt[\beta]{\theta_v} \cdot \sigma_w \tag{5}$$

$$\kappa^{\beta} = (1-n) \cdot \kappa_s^{\beta} + n \cdot S_r \cdot \kappa_w^{\beta} + n \cdot (1-S_r) = (1-n) \cdot \kappa_s^{\beta} + \theta_v \cdot (\kappa_w^{\beta} - 1) + n \tag{6}$$

Non-destructive techniques, such as TDR, permit measuring conductivity and permittivity of a given soil. Furthermore calibrated equations allow the estimation of porosity and volumetric water content (Jones et al. 2001; Noborio 2001). That is, if electrical conductivity measurements are obtained, the results may be correlated to volumetric water content; and if the dielectric permittivity is measured, two unknown parameters may be inverted for: porosity and volumetric water content. By combining Equations 5 and 6, the mass density may be estimated if the specific gravity is assumed to be known. The problem with this analysis is that the water conductivity is seldom known. Furthermore, dielectric permittivity measurements have been traditionally correlated to volumetric water content (Table 1).

Table 1. Evaluation of volumetric water content using TDR measurements

Researcher	Equation
Topp et al. (1981)	$\theta_v = -5.3 \cdot 10^{-2} + 2.92 \cdot 10^{-2} \cdot \kappa - 5.5 \cdot 10^{-4} \cdot \kappa^2 + 4.3 \cdot 10^{-6} \cdot \kappa^3$
Mixture equation ($\beta \approx 0.5$)	$\theta_v = \dfrac{\kappa^{\beta} - (1-n) \cdot \kappa_s^{\beta} - n \cdot \kappa_a^{\beta}}{\kappa_w^{\beta} - \kappa_a^{\beta}}$
Maliki et al. (1996)	$\theta_v = \dfrac{\sqrt{\kappa} - 0.819 - 0.168 \cdot \rho - 0.159 \cdot \rho^2}{7.17 + 1.18 \cdot \rho}$

Source: Topp et al. (1981); Benson and Bosscher (1999); Jones et al. (2001); Noborio (2001)

4. TDR EVALUATION OF DENSITY AND WATER CONTENT

The question is then: How does the Purdue TDR method work in the estimation of the mass density and water content? Replacing the value of the volumetric water content by $\theta_v=\rho_d/\rho_w$ w, where ρ_d is the dry density and w is the gravimetric water content, into Equation 6:

$$\kappa_{in\,situ}{}^{\beta} = \left[1-n_{in\,situ}\right]\cdot\kappa_s^{\beta} + \frac{\rho_{d\,in\,situ}}{\rho_w} w\cdot\left(\kappa_w^{\beta}-1\right)+n_{in\,situ} \quad (7)$$

$$\kappa_{compacted}{}^{\beta} = \left[1-n_{compacted}\right]\cdot\kappa_s^{\beta} + \frac{\rho_{d\,compacted}}{\rho_w} w\cdot\left(\kappa_w^{\beta}-1\right)+n_{compacted} \quad (8)$$

in these two equations, w, κ_s, κ_w, ρ_w and β are assumed to be the same for both the in situ and the compacted soils. Furthermore w, $n_{compacted}$ and $\rho_{d\ compacted}$ are known because they are measured in a modified compaction mold, therefore when $\kappa_{compacted}$ is measured, the only unknown parameter in Equation 8 is κ_s. Once this parameter is solved for, the unknown parameters in Equation 7 are $\rho_{d\ in\ situ}$ and $n_{in\ situ}$. Replacing porosity $n_{in\ situ}$ in the equation for dry mass density $\rho_{d\ in\ situ}=(1-n_{in\ situ})\cdot G_s\cdot\rho_w$:

$$\kappa_{in\,situ}{}^{\beta} = \left(1-n_{in\,situ}\right)\cdot\kappa_s^{\beta} + \frac{1-n_{in\,situ}}{\rho_w} G_s\cdot w\cdot\left(\kappa_w^{\beta}-1\right)+n_{in\,situ} \quad (9)$$

The unknown parameters in Equation 9 are the specific gravity G_s and $n_{in\ situ}$. The specific gravity varies only between 2.6 and 2.8. Then, if G_s is set at 2.7, the error in Equation 9 would be less than 3.5 %. The main assumption in this derivation is that the in situ soil and the compacted soil are the same, and that the water content does not vary throughout the testing site. Such an assumption and the fact that the soil specimens must be removed at regular intervals could limit the applicability of this methodology.

It is clear from published papers (Drnevich et al. 2003; Yu and Drnevich 2004a; 2004b), that the Purdue TDR method does not use this analysis, however this derivation shows why the methodology has a robust foundation and it sheds light to possible limitations in the estimation of mass density and water content with the Purdue TDR method.

5. ADDING INFORMATION: ELASTICS WAVE PARAMETERS

The previous analysis indicates that short of removing a soil specimen from the site under evaluation, there is not enough data to evaluate the mass density and water content from the electric properties. A promising alternative is to add independent information by using other physical means. One of these alternatives is to use elastic wave velocity. The elastic wave velocity depends on the stiffness of the soil and the mass density. Furthermore, the stiffness depends on the applied external stresses, cementation and the capillary forces (see for example Richart *et al.* 1970; Wu *et al.* 1984; Qian *et al.* 1993). In the case of freshly remolded soils near the surface (i.e., newly compacted bases) the effect of cementation and external stresses is minimum and the controlling parameter in the evaluation of the bulk stiffness is the capillary forces as represented by the degree of saturation.

To permit the evaluation of the P-wave velocity as function of the degree of saturation, Fratta et al. (2005) proposed an equation that relates the P-wave velocity to the degree of saturation (the semi-empirical model is based on the model presented by van Genuchten 1980):

$$V_P = \sqrt{\frac{\dfrac{1}{n\left(\frac{S_r}{B_w} + \frac{1-S_r}{B_a}\right) + \frac{1-n}{B_g}} + \dfrac{7}{3} G_0 \left(S_r^{\frac{m}{1-m}} - 1\right)^{\frac{1}{m}}}{\rho_w \cdot G_s (1-n) + \rho_w \cdot n \cdot S_r}} \qquad (10)$$

where B_w, B_a and B_g are the bulk modulus of the water, air and grain mineral respectively, G_0 is the shear modulus and m is an experimentally determined parameter. In Equation 10, B_w, B_a and B_g are assumed to be known and G_s is 2.7. There are three unknown parameters: n, G_0 and m. These parameters relate the variation of the skeleton shear stiffness for changing degree of saturation and they depend on the grain size distribution and texture of the soil (see for example Cho and Santamarina 2002). Figure 2 shows the variation of the P-wave velocity with volumetric water content for varying values of m.

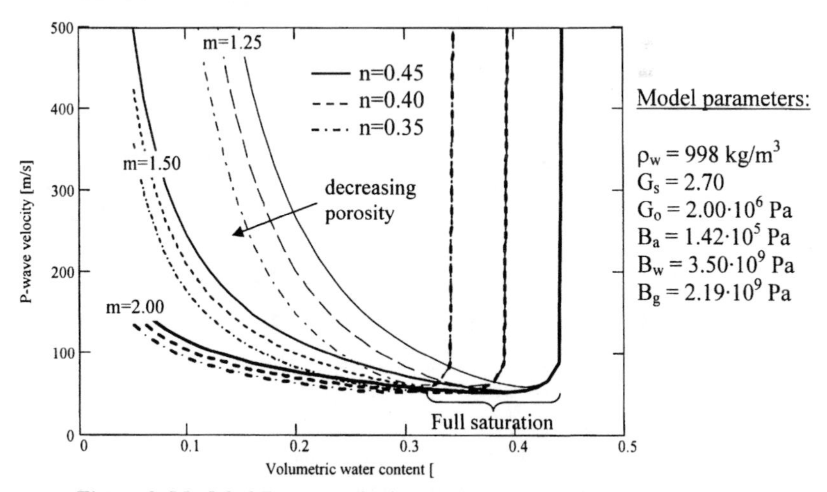

Figure 2. Modeled P-wave velocity profiles for different values of the experimentally determined parameter m and porosity.

The problem with Equation 10 is that the m parameter must be externally calibrated unless the soil can be forced to change the degree of saturation while monitoring the change in P-wave velocity. That is, if the water content and P-wave velocity are monitored as the soil saturation changes, all three parameters n, G_0 and m may be determined. To facilitate this analysis, volumetric water content θ_v replaces $n \cdot S_r$ in Equation 10:

$$V_P = \sqrt{\frac{\dfrac{1}{\frac{\theta_v}{B_w} + \frac{n-\theta_v}{B_a} + \frac{1-n}{B_g}} + \dfrac{7}{3} G_0 \left[\left(\frac{\theta_v}{n} \right)^{\frac{m}{1-m}} - 1 \right]^{\frac{1}{m}}}{\rho_w \cdot G_s (1-n) + \rho_w \cdot n \cdot S_r}} \quad (11)$$

The change in water content is monitored using TDR technology while the P-wave velocity can be monitored using piezocrystal elements. The main assumption is that as the degree of saturation increases, the porosity of the soil remains constant. This assumption limits the application of the methodology to non-expansive soils. This methodology could eliminate the need to remove a soil specimen and the calibration of the equation since it is self-calibrated.

6. COMBINED MODEL OF INTERPRETATION

The model presented in the previous section relates the P-wave velocity of soils for increasing water contents. The question is: Can the model be used to solve for the parameters needed in the inversion of the mass density and water content? To answer this question, a numerical analysis is proposed.

The model presented in Equation 11 is used to generate synthetic data. To create a more realistic interpretation model, white random noise is added to the velocity (±5%) and to the volumetric water content (±2%). These errors are the commonly accepted in the measurement of wave velocity and water content (Hagedoorn 1964; Noborio 2001). Table 2 presents the synthetic values with and without noise and Figure 3 shows the P-wave velocity versus volumetric water content as the soil is inundated.

To be able to solve for the unknown parameters, the methodology constrains the selected inverted parameters by:

Minimizing the square of the residual error:

- $$\text{Error} = \frac{1}{N} \sqrt{\sum_{i=1}^{N} \left(V_{P_i}^{<measured>} - V_{P_i}^{<calculated>} \right)^2} \quad (12)$$

  where $V_{P_i}^{<measured>}$ is the synthetic-generated measured P-wave velocity at volumetric water content θ_{vi}, $V_{P_i}^{<calculated>}$ is the model-estimated P-wave velocity (using Equation 11), and N is the number of measurements at increasing water content.

- The value of porosity n should be larger than the maximum measured value of volumetric water content.
- The value of parameter m should be greater than one.
- The ratio between P- and S-wave velocities should be approximately equal to 1.5 for a Poisson's ratio $\nu = 0.15$.

These added constraints help in the convergence of the solution. Results are presented in Figures 4 and 5 and in Table 2. The inverted parameters include m, G_o, and n. The porosity n and the initial water content are then used to calculate the density, dry density, and gravimetric water content using the following equations:

$$\rho = 2.7 \cdot (1-n) \cdot \rho_w + \theta_v \cdot \rho_w \tag{13}$$

$$\rho_d = 2.7 \cdot (1-n) \cdot \rho_w \tag{14}$$

$$w = \frac{\rho}{\rho_d} - 1 \tag{15}$$

Table 2. Synthetic values, noisy data and inverted parameters

#	Specific gravity	Assumed Porosity	Assumed ini. water content	Ini. vol. water content	Ini. noisy vol. water content	Ini. noisy synthetic velocity	Inverted porosity	Inverted ini. water content
1	2.6	0.35	0.10	0.169	0.166	107.0 m/s	0.346	0.094
2	2.6	0.40	0.14	0.218	0.215	128.9 m/s	0.388	0.130
3	2.7	0.37	0.12	0.204	0.204	191.4 m/s	0.374	0.121
4	2.7	0.42	0.16	0.251	0.252	144.4 m/s	0.414	0.159
5	2.8	0.40	0.14	0.235	0.231	81.3 m/s	0.398	0.142
6	2.8	0.45	0.18	0.277	0.280	145.8 m/s	0.451	0.189

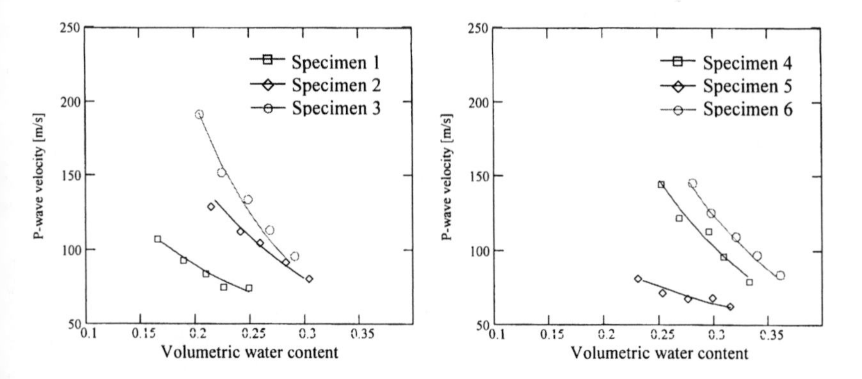

Figure 3. Modeled data with (symbols) and without (lines) uniform random noise in both the volumetric water content and P-wave velocity.

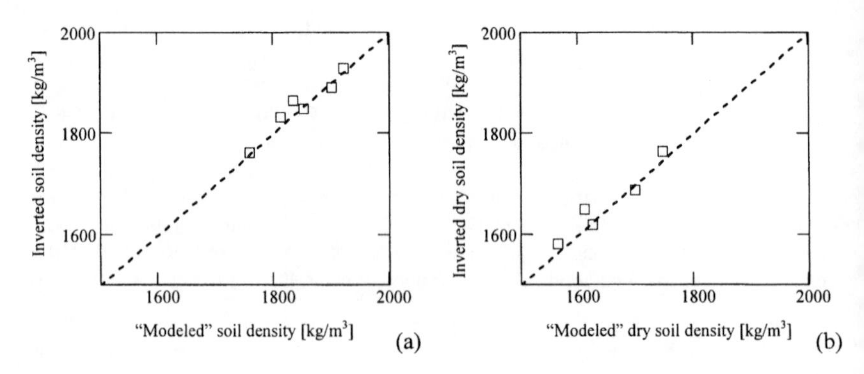

Figure 4. Synthetic data versus inverted parameters: (a) soil density and (b) dry density.

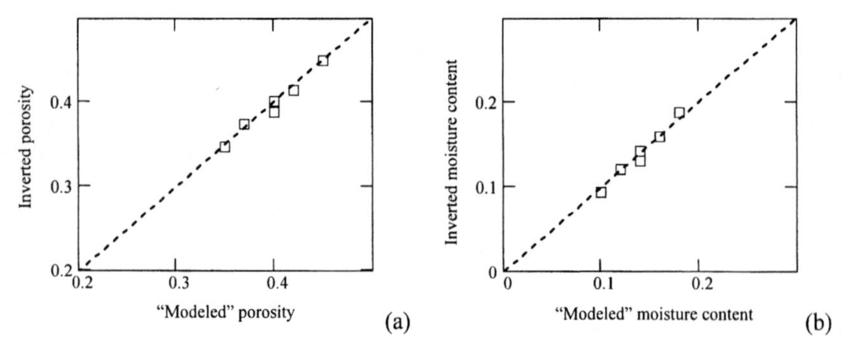

Figure 5. Synthetic data versus inverted parameter: (a) porosity and (b) gravimetric water content.

However, other constraints must be implemented to be able to implement this methodology in the field as the convergence of the error function (Equation 12) is not very steep and problems with non-unique solutions may be found. For example, Figure 6 shows ranges of error functions that yield similar results for different combination of unknown shear stiffness and porosity values.

7. CONCLUSIONS

Evaluation of the water content and mass density in soils using new non-destructive methods must be based on solid physical principles in order to properly estimate the required parameters. This paper presents a combined methodology that uses both electromagnetic and elastic wave parameters to yield enough information to solve for soil phase properties. A simple numerical analysis of the convergence of the solution shows that the proposed methodology can be used to estimate the mass density and water content when both the dielectrical permittivity and P-wave velocity are measured while the water content is increased in the soils. A solution is obtained even when simulated measurement errors are presented both in the evaluation of volumetric water content and P-wave velocity. However, the analyst should incorporate other physically meaningful constraints to facilitate the convergence of the solution for field applications.

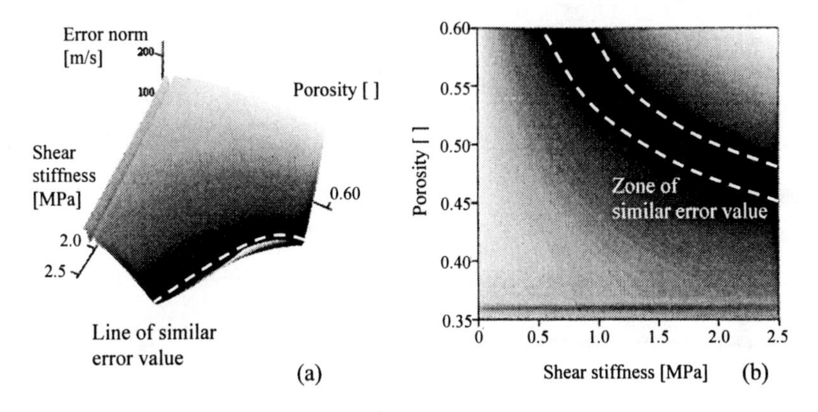

Figure 6. The error norm shows that even for this simulated problem the convergence is not guarantee: (a) Surface plot and (b) contour plot. The error norm is almost flat along a large area of the error norm plot (the darker region in both plots).

8. ACKNOWLEDGEMENTS

The authors gratefully acknowledge the financial support provided by the US Army-SBIR Program (Topic No. A012-0225). Thanks are also due to Mr. Steven Trautwein, our partner in the US Army SBIR project. Finally, the authors acknowledge the careful comments of an anonymous reviewer.

9. REFERENCES

Abu-Farsakh, M. Y., Alshibli, K. A., Nazzal, M. and Seyman, E. (2004). "Evaluating the Stiffness of Highway Materials from the Geogauge Device". *Geo Jordan 2004: Advances in Geotechnical Engineering with Emphasis on Dams, Highway Materials, and Soil Improvement*. ASCE Press. pp. 287-298.

ASTM D1556-00 (2005). *Standard Test Method for Density and Unit Weight of Soil in Place by the Sand-Cone Method.* ASTM International. West Conshohocken, PA.

ASTM D2167-94 (2001). *Standard Test Method for Density and Unit Weight of Soil in Place by the Rubber Balloon Method.* ASTM International. West Conshohocken, PA.

ASTM D3017-04 (2004). *Standard Test Method for Water Content of Soil and Rock in Place by Nuclear Methods (Shallow Depth).* ASTM International. West Conshohocken, PA.

ASTM D6758-02 (2005). *Standard Test Method for Measuring Stiffness and Apparent Modulus of Soil and Soil-Aggregate In-Place by an Electro-Mechanical Method).* ASTM International. West Conshohocken, PA.

ASTM D6780-02 (2005). *Standard Test Method for Water Content and Density of Soil in Place by Time Domain Reflectometry (TDR).* ASTM International. West Conshohocken, PA.

Benson, C. H. and Bosscher, P. J. (1999). "Time-Domain Refelctormetry (TDR) in Geotechnics: A Review". *Nondestructive and Automated Testing for Soil and Rock Properties.* ASTM STP 1350. Edited by W. A. Marr and C. E, Fairhust. American Society of Testing and Materials. West Conshohocken, PA. pp.113-136.

Cho, G. C., and J. C. Santamarina (2001), Unsaturated Particulate Materials - Particle Level Studies, *Journal of Geotechnical and Environmental Engineering*, Vol. 127, No. 1, pp. 84-96.

Daniel, D. E., and Benson, C. H. (1990). "Water Content-Density Criteria for Compacted Soil Liners". *Journal of Geotechnical Engineering*. Vol. 116, No. 12, pp. 1811-1830.

Drnevich, V. P., Lin, C. P., Yi, Q., and Lovell, J. (2001a). "Final Report: SPR-2201, Real-Time Determination of Soil Type, Water Content, and Density Using Electromagnetics". *FHWA/JTRP*, IN-2000/20, File No. 6-6-20, 320 pages.

Fratta, D., Alshibli, K. A., Tanner, W. M., and Roussel, L. (2005). "Combined TDR and P-wave Velocity Measurements for the Determination of In Situ Soil Density – Experimental Study". *ASTM Geotechnical Testing Journal* (Accepted for Publication – November 2005).

Hagedoorn, J. G. (1964). "The Elusive First Arrival". *Geophysics*, Vol. 29, No. 5, pp. 806-813.

Jones, S. B., Wairth, J. M. and Or, D. (2001). "Time Domain Reflectometry Measurement Principles and Applications". *Hydrological Processes*. Vol. 16. pp. 141-153.

Lin, C.-P., Drnevich, V. P., Feng, W. and Deschamps, R. J. (2000). "Time Domain Reflectometry for Compaction Quality Control". *Use of Geophysical Methods in Construction,* Edited by S. Nazarian and J. Diehl, Geophysical Special Publication 108, ASCE Press, pp. 15-34.

Noborio, K. (2001). "Measurement of Soil Water Content and Electrical Conductivity by Time Domain Reflectometry: A Review". *Computers and Electronics in Agriculture*. Vol. 31. pp. 213-237.

Qian, X., Gray, D. H., and Woods, R. D. (1993). "Voids and Granulometry: Effects on Shear Modulus of Unsaturated Sands". *Journal of Geotechnical Engineering*, Vol. 119, No. 2, 1993, pp. 296-314.

Richart, F. E., Hall, J. R., and Wood, R. D. (1970). *Vibrations of Soils and Foundations*. Prentice-Hall. Englewood, NJ, 1970.

Santamarina, J. C. in collaboration with Klein, K. A. and Fam, M. A. (2001). *Soils and Waves*. John Wiley and Sons. Chichester, UK. 488 pages.

Sawangsuriya, A., Bosscher, P. J., and Edil, T. B. (2003). "Laboratory Evalaution of the Soil Stiffness Gauge". *Transportation Research Record 1808*. pp. 30-37.

Sawangsuriya, A., Edil, T. B., and Bosscher, P. J. (2004). Assessing small-strain stiffness of soils using the soil stiffness gauge, *Proceedings of the 15th Southeast Asian Geotechnical Society Conference*, Bangkok, Thailand, pp.101-106.

Siddiqui, S. I. and V. P. Drnevich (1995). "A New Method of Measuring Density and Moisture Content of Soil Using the Technique of Time Domain Reflectometry". *Report No: FHWA/IN/JTRP-95/9*, Joint Transportation Research Program, Indiana Department of Transportation-Purdue University, February 1995, 271 pages.

Topp, G. C., Davis, J. L. and Annan, A. P. (1980). "Electromagnetic Determination of Soil Water Content: Measurements in Coaxial Transmission Lines". *Water Resources Research*. Vol. 16, No. 3, pp. 574-582.

TransTech (2005). TransTech Company Web Site. URL: www.transtechsys.com

van Genuchten, M. Th. (1980). "A Closed-form Equation for Predicting the Hydraulic Conductivity of Unsaturated Soils". *Soil Science Society Science of America Journal*. Vol. 44, pp. 892-898.

Wu, S., Gray, D. H., and Richart, F. E. (1984). "Capillary Effects on Dynamic Modulus of Sands and Silts". *Journal of Geotechnical Engineering*, Vol. 110, No. 9, pp. 1188-1203.

Yu, X. and Drnevich, V. P. (2004a). "Time Domain Reflectometry for Compaction Control of Stabilized Soils". *Transportation Research Record No. 1868*. pp. 14-22.

Yu, X. and Drnevich, V. P. (2004b). "Soil Water Content and Dry Density by Time Domain Reflectometry". *Journal of Geotechnical and Geoenvironmental Engineering*. Vol. 130, No. 9, pp. 922-934.

STRESS REDUCTION BY ULTRA-LIGHTWEIGHT GEOFOAM FOR HIGH FILL CULVERT: NUMERICAL ANALYSIS

Liecheng Sun[1], P.E., Tommy Hopkins[2], P.E., P.G.
and Tony Beckham[3], P.G.

ABSTRACT: The study of earth pressure distribution on buried structures has a great practical importance in constructing highway embankments above pipes and culverts. Based on Spangler's research, the supporting strength of a conduit depends primarily on three factors: 1. the inherent strength of the conduit; 2. the distribution of the vertical load and bottom reaction; and, 3. the magnitude and distribution of lateral earth pressures which act against the sides of the structure. Considering high fills above them and high earth pressures they may experience, rigid culverts are usually used underneath highway embankments. To reduce high vertical earth pressures acting on buried structure, ultra-lightweight Geofoam was placed above a culvert in the field, in Russell County, KY. Before construction began, numerical analysis using FLAC 4.00 (Fast Lagrangian Analysis of Continua) had been performed to predict stresses on the culvert. Results of the analysis show that Geofoam has a great effect in reducing vertical stresses above and below the culvert. There were areas of high stress concentrations at the top and bottom of the concrete culvert if no Geofoam was placed above the culvert. After placing Geofoam above the culvert, the concentrated stress at the top can be reduced to 28 percent of the stress without Geofoam. The high stress at the bottom of culvert can be reduced to 42 percent of the stress without Geofoam. Stresses on the two sidewalls of the culvert were observed to have no significant change in values with and without Geofoam.

INTRODUCTION

Construction of highway embankments above highway pipes and culverts has a great practical significance because of stresses imposed by the fill on the buried

[1]Research Engineer, University of Kentucky Transportation Center, University of Kentucky, Lexington, KY 40506-0281, lsun00@uky.edu
[2]Head of Geotechnology, University of Kentucky Transportation Center, University of Kentucky, Lexington, KY 40506-0281
[3]Research Geologist, University of Kentucky Transportation Center, University of Kentucky, Lexington, KY 40506-0281

structure. The relative stiffness of the culvert and fill controls the magnitude and distribution of earth pressures on the buried structure. The vertical earth pressure on a flexible culvert, or a culvert with a yielding foundation, is less than the weight of the soil above the culvert due to positive arching. However, the vertical earth pressure on a rigid culvert with a non-yielding foundation is greater than the weight of the soil above the structure because of negative arching. Experiments by Marston (Spangler, 1958) showed that loads on rigid embankment culverts were some 90 to 95 percent greater than the weight of the soil directly above the structure. In model tests performed by Hoeg (1968), the crown pressure was about 1.5 times the applied surcharge. Penman et al. (1975) measured the earth pressure on a rigid reinforced concrete earth pressure below 174 feet of rock fill and found that the vertical earth pressure on the culvert crown was about 2 times the overburden stress due to the fill above the top of the culvert.

Based on Spangler's research, the supporting strength of a conduit depends primarily on three factors: first, the inherent strength of the conduit; second, the distribution of the vertical load and the bottom reaction; and third, the magnitude and distribution of lateral earth pressures, which may act against the sides of the structure. The last two factors are greatly influenced by the character of the bedding on which the culvert is founded and by the backfilling against the sides. Considering the high fills above them and the high earth pressure they may experience, rigid culverts are usually used underneath highway embankments. To reduce large vertical earth pressures on buried structures, the imperfect ditch method of construction introduced by Marston (Handy and Spangler, 1973) can be used. This method has considerable merit from the standpoint of minimizing the load on a culvert under an embankment. Figure 1 shows a sketch of the traditional installation of the imperfect ditch culvert.

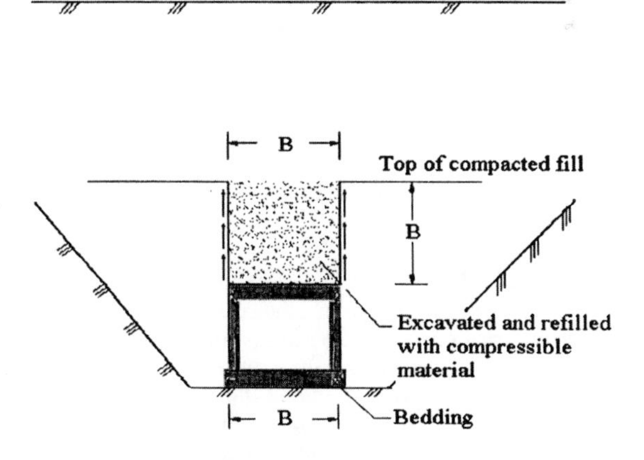

FIG. 1. Imperfect ditch culvert traditional installation

This method involves installing a compressible layer above the culvert within the backfill. In field construction, the culvert is first installed as a positive projecting conduit and then surrounded by thoroughly compacted backfill. Next, a trench is dug in the compacted soil directly above the culvert. The trench is backfilled with compressible material, or organic fill, creating a soft zone. When the embankment is constructed, the soft zone compresses more than its surrounding fill, and thus positive arching is induced above the culvert. Traditionally, organic materials such as baled straw, leaves, old tires (used in France), or compressible soil, have been used. Very little quantifiable data is available about the stress-strain properties of the soft organic materials. Also, the long-term stability and performance of the organic material is also questionable.

Expanded polystyrene (EPS, or Geofoam) can be used as the compressible material to promote positive arching (Vaslestad et al., 1993). EPS has low stiffness and exhibits the desirable elastic-plastic behavior. An unconfined compressive strength test was conducted on EPS by University of Kentucky Transportation Research Center and the result showed that its stress-strain behavior is very similar to an ideal elastic-plastic material (Figure 2). The maximum compressive strength of EPS obtained from the test is about 3.0 ksf, and with the Young's modulus in the linear range is 133 ksf.

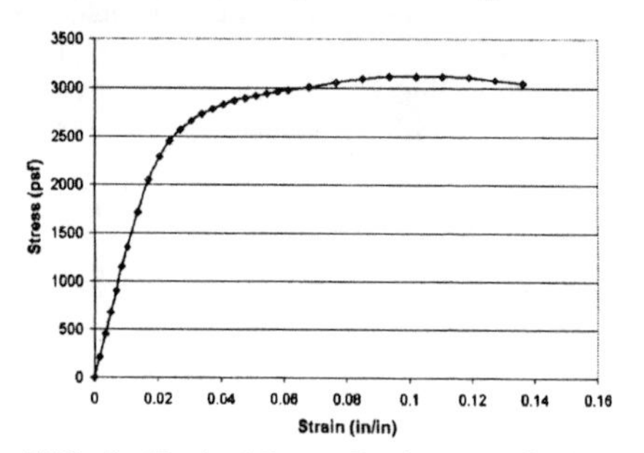

FIG. 2. Typical Stress-Strain curve for EPS

SITE DESCRIPTION

A culvert located on the Jamestown Bypass (US 127) in Russell County, Kentucky was selected for theoretical analyses and eventually instrumentation. Rock cores taken from this location revealed fossiliferous limestone with many shale laminations on which the culvert will be constructed. The culvert is a cast-in-place box culvert. The inner width of the structure is 9 feet and the wall thickness is 1 foot. The inner height is 8 feet and the ceiling thickness is 2 feet and 1 inch. The bottom thickness of the slab is 2 feet and 2 inches. It is continuously placed on an unyielding foundation along a total length of 370 feet, and crosses a valley beneath an embankment of compacted backfill up to 54 feet above the culvert.

To investigate different pressures on the culvert due to placement of EPS (Geofoam), three different sections were selected from the same culvert. On the first section, 2 feet of EPS was placed above the culvert. The width of EPS was the same as the top of the culvert (11 feet) as shown in Figure 3. On the second section, EPS was placed above the culvert directly at 2 feet thickness and a width of 16 feet, which is 1.5 times the culvert width as shown in Figure 4. The length of both sections is 20 feet. The EPS sections placed where the fill is highest, 54 feet. The third section was a conventional one, which is used as a reference section for the other two EPS sections. The three sections were instrumented to measure the stresses on the top and sides of culvert. The strain of the top slab were also measured. Three "sister" reinforcing steel bars containing strain gages were placed in the culvert during construction. Twelve earth pressure cells placed on the top and one side of the structure.

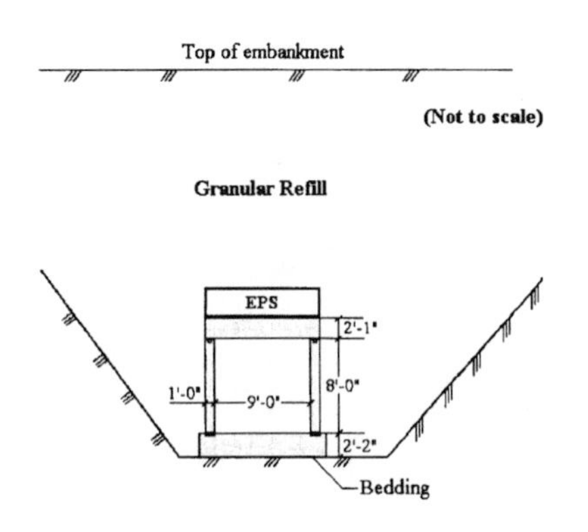

FIG. 3. Same width EPS on culvert

NUMERICAL ANALYSIS USING FLAC

The purpose of this analysis is to investigate the pressure changes due to the placement of EPS on top of the culvert using the two-dimensional finite difference program FLAC (Version 4.0, Itasca). A set of computer runs were used to identify the optimal situation as a function of the EPS size and position. Numerical analyses were also conducted to investigate the effect of using different combinations of elastic modulus, Poisson's ratio, cohesion, and angle of internal friction of the backfill.

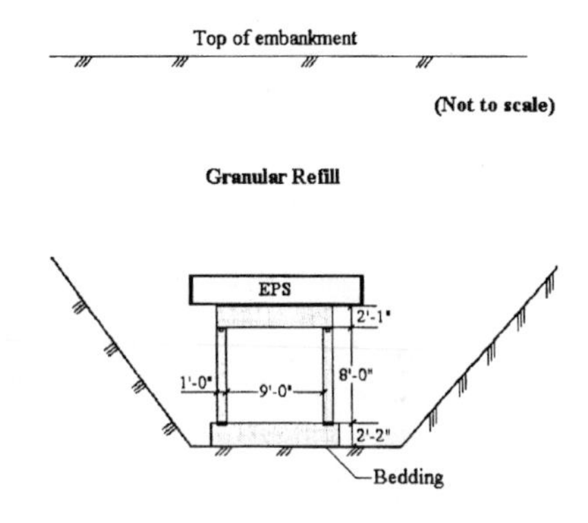

FIG. 4. 1.5 times culvert width EPS

Numerical Model and Properties of Materials

Solving a problem using FLAC involves thousands of iterations. To speed up the iteration calculation, half space has been considered for this symmetrical problem (Figure 5). The culvert was treated as a beam element with hinges on the upper and bottom corners. Interface elements were used between the culvert and soils or EPS.

The properties of materials, except EPS, used in this analyses were based on data available in the report by the Commonwealth of Kentucky Transportation Cabinet, Department of Highways, Division of Bridge Design. They represent typical values used in design practice.

The backfill soil was modeled as a cohesionless material using FLAC plastic constitutive model that corresponds to a Mohr-Coulomb failure criterion.

Bedrock and concrete were modeled as linear-elastic materials. Considering model availability in FLAC, EPS was also modeled as a linear-elastic material. In this imperfect ditch approach, this model will create more conservative results. The specific material properties used in the FLAC software are listed in Table 1.

TABLE 1. Material Properties

Material (1)	Elastic Modulus E (psf) (2)	Poisson's Ratio υ (3)	Mass Density (pcf) (4)	Friction Angle ϕ (5)
Concrete	543×10^6	0.35	156	
EPS	0.133×10^6	0.1	1.35	
Sandy Gravel	4.177×10^6	0.35	120	34°
Bedrock	108×10^6	0.25	120	

Calibration of the Numerical Model

The roughly described properties used in job site for backfill material yield some uncertain factors for numerical analysis. The varied sizes of EPS make the analyses more complicated. Based on the original design conditions, the numerical model was calibrated by adjusting interface parameters between the culvert and backfill, and by trying different combinations of elastic modulus, Poisson's ratio, and angle of internal friction of the backfill. The maximum pressure and total vertical load on top of the culvert obtained from numerical modeling are adjusted to the numbers showed in the report by the Commonwealth of Kentucky Transportation Cabinet, Department of Highways, Division of Bridge Design (Figure 6).

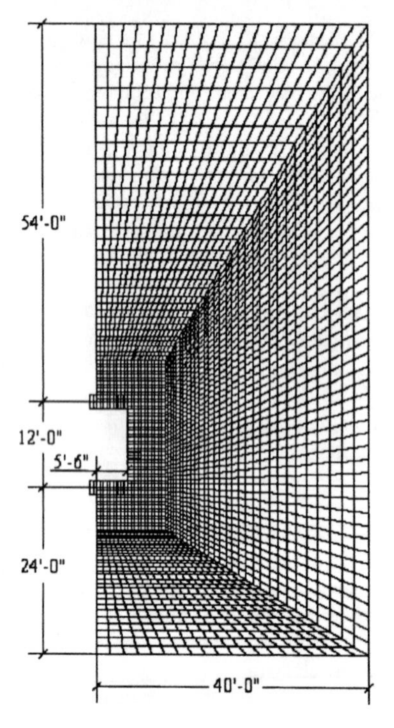

FIG. 5. Model mesh

Analyses of Stresses on Culvert Using Different Sizes of EPS

To investigate the effects on the earth pressure in a backfill using the imperfect ditch method, EPS is placed above the culvert directly. Two sets of parametric studies were used to investigate stress distributions with different combinations of elastic modulus, Poisson's ratio, cohesion, and friction angle for backfill under two different sizes of EPS (Figures 3 and 4). Typical results, corresponding to design loads, are shown in Figures 7 through 9.

Maximum Vertical Distributed Load On Top of Culvert:

Designed	From FLAC
15.3306 K/Ft	15.35 K/Ft

$k1*k2*k3 = 2.355$

$q_{max} = 2.355*\gamma*h$

(Based on research Report UKTRP-84-22)

Total Vertical Load On Top of Culvert:

Designed	From FLAC
52.91 Kips	53.00 Kips

FIG. 6. Calibration of the numerical model

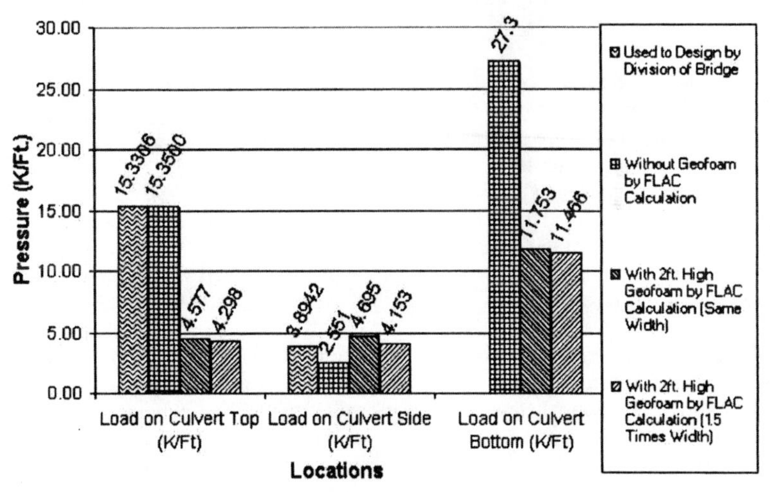

FIG. 7. Comparison of maximum pressures on culvert with and without EPS

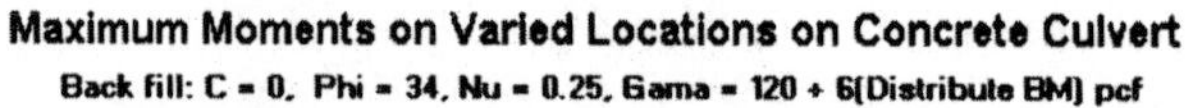

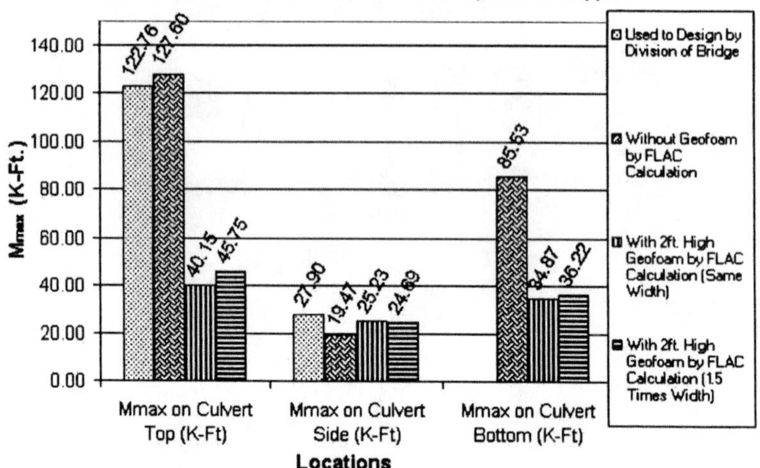

FIG. 8. Comparison of maximum moments on culvert with and without EPS

The numerical results show that the maximum pressure at the top of culvert, with EPS width 1.5 times the culvert width, is reduced to 4.298 kips/ft, which is 28 percent of the maximum pressure without EPS. When the width of EPS equals the width of culvert, the maximum pressure at the top of culvert is reduced to 4.577 kips/ft, which is 30 percent of the maximum pressure without EPS (Figure 7). The maximum moment on the top of culvert is decreased to 40.15 kip-ft/ft, which is 32 percent of the maximum moment without EPS (Figure 8). The interesting point is that the maximum moment is smaller when EPS width is the same as the culvert width (Figure 8). The possible reason to explain this result is that narrower EPS creates a larger arching effect.

The maximum pressure at the bottom of culvert reduced to 11.466 kips/ft, when the EPS width is 1.5 times the culvert width, which is 42 percent of the pressure without EPS. In the situation where width of EPS equals to width of culvert, the maximum pressure at the bottom of culvert is reduced to 11.753 kips/ft, which is 43 percent of the maximum pressure without EPS (Figure 7). The maximum moment on the bottom of the culvert is decreased to 34.87 kip-ft/ft, when the width of EPS equals the width of the culvert, which is 41 percent of the maximum moment without EPS (Figure 8).

The maximum pressure on the sidewall of culvert is increased to 4.695 kips/ft, which is 84 percent more than the pressure without EPS, when EPS width equals culvert width. In the situation where EPS width is 1.5 times the culvert width, the maximum pressure on the sidewall of culvert is increased to 4.153 kips/ft, which is 63 percent more than the maximum pressure without EPS (Figure 7). However,

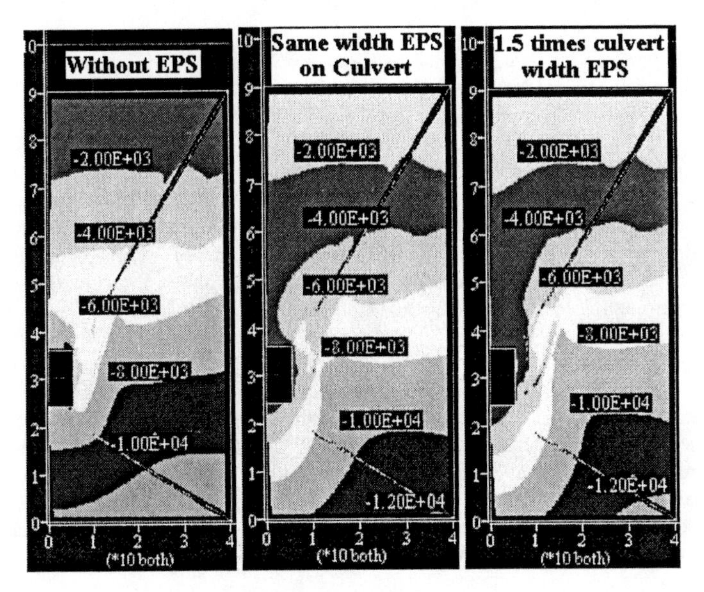

FIG. 9. Contours of maximum principal stress with and without EPS on the top of culvert (psf)

compared to the design load used by the Kentucky Transportation Cabinet, those values are increased 21 percent and 6.6 percent for the case of same EPS width as culvert width and EPS width equal 1.5 times the culvert width, respectively. The maximum moment on the sidewall of the culvert was 30 percent more when the widths of EPS and the culvert are the same. But, that value is still 9.6 percent lower than the design value used by the Kentucky Transportation Cabinet (Figure 8).

The stress reduction is also observed from contours of maximum principal stress as shown in Figure 9. Comparing stress contours between with and without geofoam, the lower stress zone is extended to culvert top, side, and bottom for the situations with geofoams. The wider the geofoam, the deeper the lower stress area is projected in this specific case.

CONCLUSIONS AND DISCUSSIONS

Results of the numerical analysis showed that the EPS has a great effect in reducing the vertical soil pressures above and below a culvert. When EPS is not placed above the culvert, areas of high stress concentrations occur at the top and bottom of the concrete culvert. By placing EPS above the culvert, the concentrated stress at the top of the culvert can be reduced to 28 percent of the concentrated stress without EPS. The highest stress at the bottom of culvert can be reduced to 42 percent of the highest stress without EPS. Whether EPS is used or not used, the model analysis showed that the maximum moment acting on the sidewall does not change significantly. Although the maximum moment acting on the sidewall is higher when EPS is used, the value is still below the design value used by the Kentucky Transportation Cabinet.

The linear-elastic model was used to simulate the EPS stress-strain behavior in this numerical analysis. As pointed out earlier, the EPS exhibits desirable elastic-plastic behavior during compression (Figure 2). The EPS creates larger deformation, which develops higher positive arching effect under elasto-plastic model especially when stress on EPS is beyond elastic range. The positive arching will further reduce the pressure on the culvert. The ground water table is an important factor but was not considered in the analysis due to the lack of field information. Considering the high fills above the culvert, the ground water table may be above the culvert and have some non-negligible effect on the stress distribution around the culvert.

ACKNOWLEDGMENTS

Financial support for this project was provided by the Kentucky Transportation Cabinet. The authors acknowledge the Kentucky Transportation Cabinet, Department of Highways, Division of Bridge Design for providing a detailed initial design report. Special thanks to Jim King and Allan Frank, Division of Bridge Design, Mark Robertson, Resident Engineer, and Larry Kerr, Construction Branch Manager, District 8, for their cooperation.

REFERENCES

Allen, D. and Meade, B., (1984) Analysis of Loads and Settlements for Reinforced Concrete Culvert, *Research Report*, UKTRP-84-22.

Handy, R. L., and Spangler, G., (1973) Loads on Underground Conduits, *Soil Engineering*, Third Edition, 1973.

Hoeg, K., (1968), Stresses Against underground Structural Cylinders, *Journal of the Soil Mechanics and foundation Division*, ASCE, Volume 94, No. SM4, 833 – 858.

Penman, A.D.M., Charles, J. A., and Nash, J.K., and Humphreys, J.D., (1975), Performance of Culvert Under Winscar Dam, *Geotechnique*, Volume 25, No. 4, 713 - 730.

Spangler, M. G., (1958), A Practical Application of the Imperfect Ditch Method of Construction, *Proceedings of Highway Research Board*, Volume 37.

Vaslestad, J., Johansen, T. H., and Holm, W. (1993) "Load Reduction on Rigid Culverts Beneath High Fills". *Transportation Research Record*, 1415, 58 – 68.

APPENDIX I. CONVERSION TO SI UNITS

Feet (ft) X 0.305 = meter (m)
Pounds per cubic foot (pcf) X 16.018 = kilogram on per cubic meter (kN/m^3)
Pounds per square foot (psf) X 47.881 = pascal (Pa)
Kip X 4.4482 = kilonewton (kN)
Kip-ft (K-ft) X 1.3558 = kilonewton-meter (kN-m)

Subject Index

Page number refers to first page of paper

Author Index

Page number refers to first page of paper